T0191860

The Society of Fire Protection Engineers Series

Series Editor
Chris Jelenewicz
Society of Fire Protection Engineers
Gaithersburg, MD, USA

The Society of Fire Protection Engineers Series provides rapid dissemination of the most recent and advanced work in fire protection engineering, fire science, and the social/human dimensions of fire.

The Series publishes outstanding, high-level research monographs, professional volumes, contributed collections, and textbooks.

Society of Fire Protection Engineers

SFPE Guide to Fire Risk Assessment

SFPE Task Group on Fire Risk Assessment

Society of Fire Protection Engineers
Gaithersburg, MD, USA

ISSN 2731-3638 ISSN 2731-3646 (electronic)
The Society of Fire Protection Engineers Series
ISBN 978-3-031-17702-6 ISBN 978-3-031-17700-2 (eBook)
https://doi.org/10.1007/978-3-031-17700-2

This Springer imprint is published by the registered company Springer Nature Switzerland AG
The registered company address is: Gewerbestrasse 11, 6330 Cham, Switzerland

SFPE Task Group on Fire Risk Assessment

CHAIR

Francisco Joglar, PE, PhD
Jensen Hughes

Vice-Chair

Victor Ontiveros, PhD
Jensen Hughes

MEMBERS

Name	Employer
Brian Ashe	ABCB
Greg Baker, PhD, CPEng, FIEAust	Fire Research Group Ltd
Jason Butler, PE	Encompassed Engineering, LLC
David Charters, PhD, CEng	Risktec
Albert Condello	AC Fire & Safety Consultants LLC
Luca Fiorentini	TECSA
Kevin Frank, PhD	BRANZ, Ltd
Håkan Frantzich, PhD, FSFPE	Lund University
Ashley Lindeman	Electric Power Institute
Vladimír Mózer, PhD	Czech Technical University in Prague
Tod Ossmann, PE, FSFPE	Willis Towers Watson
Rob Plonski, PE	US Department of Energy, National Nuclear Security Administration
Stephanie Poole, PE	Poole Fire Protection
Ai Sekizawa, PhD, FSFPE	Tokyo University of Science
John Utstrand	COWI
Armin Wolski, PhD, FSFPE	Reax Engineering

STAFF

Chris Jelenewicz, PE, FSFPE
SFPE

Acknowledgments

The SFPE Task Group on Fire Risk Assessment would like to thank the following individuals for their assistance in developing this guide: Austin Guerrazzi, SFPE, Daniel Morassutti, SFPE and Madison West, SFPE.

Contents

The original version of this book was revised. A correction is available at https://doi.org/10.1007/978-3-031-17700-2_16

Chapter 1
Introduction

Understanding the topic of fire risk assessment is essential in the practice of fire protection engineering. As part of fire protection engineering, risk assessment can be used as a comprehensive approach that integrates the fire safety factors that are generally used to evaluate potential strategies for an application. The aim of conducting a fire risk assessment is to gain insight into and characterize fire-related risks to better inform the wide range of decisions that should be made concerning a building, facility, or process as part of the design, construction, or operation [1]. As a result, implementing a fire risk assessment can lead to a safer, more efficient, and cost-effective design. Moreover, risk assessment can also be an essential tool used by fire engineers when implementing a performance-based strategy as it addresses explicitly unique aspects or uses. When used as part of a performance-based design, it can also provide a basis for developing and selecting alternative fire engineering options based on the project's needs (e.g., if the code-prescribed solution does not meet the stakeholders' needs) [2].

In 2006, the Society of Fire Protection Engineers (SFPE) published the first edition of *The SFPE Guide to Fire Risk Assessment* [3] This guide became a popular resource, providing a concise overview and summary of various topics that an engineer who is asked to conduct a fire risk assessment as part of a fire engineering design should consider. Specifically, the purpose of this document was to guide the use of risk assessment methodologies in the design and assessment of a building, facility, or process fire safety.

Research and practical experience related to fire risk assessment advanced substantially over the subsequent decade. SFPE has been monitoring this progress, and the SFPE Task Group on Fire Risk Assessment has developed and extended the guide, which has resulted in the 2023 *Second Edition, SFPE Guide to Fire Risk Assessment*. The SFPE Task Group developed this edition of the guide with significant and extensive global input from individuals with expertise in risk assessment. It builds on the topics in the first edition. It covers additional subjects, such as a broader discussion of identifying fire hazards and fire scenarios and a detailed quantitative and qualitative risk estimation methodology. It also includes a more

SFPE Guide to Fire Risk Assessment, The Society of Fire Protection Engineers Series, https://doi.org/10.1007/978-3-031-17700-2_1

comprehensive discussion of additional items in the risk management process, such as risk communication, residual risk management, and risk monitoring.

This edition of the guide also considers the requirements documented in standards governing the development and maintenance of a fire risk assessment. Examples of these standards published by global organizations include:

- *NFPA 551, Guide for the Evaluation of Fire Risk Assessments*. This standard describes various types and properties of fire risk assessment methods [4].
- *Part 4 of ASME/ANS RA-S–2008, Standard for Level 1/Large Early Release Frequency Probabilistic Risk Assessment for Nuclear Power Plant Applications.* This standard provides specific requirements for a quantitative fire risk assessment developed for commercial nuclear power plant applications [5].
- *ISO 16732-1:2012, Fire Safety Engineering – Fire Risk Assessment.* This standard provides the conceptual basis for fire risk assessment by stating the principles underlying the quantification and interpretation of fire-related risk [6].

Applicable requirements in a fire risk assessment standard governing a specific project should be met to ensure its technical quality in support of the decision-making and regulatory process.

The technical advances that support this new edition of the guide have been reported in numerous scientific and research publications, not necessarily written for and available to most individuals or organizations. The Task Group has distilled the most relevant and helpful information into a document that considers the key factors and considerations related to using risk assessment methodologies to design or evaluate buildings, facilities, or processes. Specifically, the guide addresses the following technical areas to the extent practical, focusing on expanding the application of fire risk assessment:

- Acceptable or tolerable risk criteria: A given industry may not have a criterion for risk acceptability agreed among stakeholders. This lack of criteria may discourage or prevent the use of fire risk assessment as a performance-based tool. This edition of the guide recommends a risk matrix that can be used as a criterion for risk acceptability and describes similar matrices already accepted and used in other industries. The guide also discusses the ALARP approach as a strategy for risk reduction that may not be based on predetermined acceptability criteria.
- Perhaps the most significant technical challenge in developing a fire risk assessment is the lack of applicable data to support the development of input values. Although this guide cannot explicitly address this challenge due to the increasing spectrum of potential applications of fire risk assessment, it describes technical approaches for managing it. For example, this guide provides a systematic approach for handling sensitivity and uncertainty analysis, which can assist in determining the impact of governing assumptions and uncertain input values.
- The guide describes the full scope and comprehensive qualitative and quantitative fire risk assessment examples to demonstrate the process and principles. These examples provide step-by-step guidance on treating most of the factors influencing the fire risk profile in a facility.

- Finally, the guide stresses the connection between code compliance (i.e., deterministic analysis) and fire risk assessment. The process described for developing a fire risk assessment includes the explicit treatment of regulatory requirements, design, and maintenance practices often reflected in applicable codes so that the analysis reflects the fire protection features, procedures, and practices governing the facility.

This revised *SFPE Engineering Guide to Fire Risk Assessment* aims to provide a common introduction to this field for the broad fire safety community, including fire protection engineers, design professionals, and code authorities.

In this document, the term “fire protection engineer” should be viewed as synonymous with the terms “fire safety engineer” and “fire engineer.” These terms apply to a person who applies engineering principles to prevent and mitigate the unwanted impact of fire. For practical purposes, only the term fire protection engineer is used in the remainder of the document.

1.1 Purpose

The purpose of this guide is to provide guidance for the following:

- Development, selection, and use of fire risk assessment methodologies for the design and operation of buildings, facilities, or processes.
- Addressing fire risk acceptability.
- The role of fire risk assessment in the fire safety design process.
- The role of fire risk assessment in operational fire safety and risk management.
- Communicating and monitoring fire risk in the design and operation of buildings, facilities, and processes.

Although written generally for evaluating the risk of fire scenarios, the process could apply to related hazards such as explosions, arson events, when appropriately treated from the perspective of characterizing their corresponding frequency of occurrence and consequences.

1.2 Scope

The *SFPE Engineering Guide to Fire Risk Assessment* describes a recommended process for developing a fire risk assessment for buildings, facilities, or processes in the design or operational stages. The guide also provides references to relevant information associated with this topic.

1.3 Limitations

This guide does not provide specific data or acceptance criteria supporting a fire risk assessment in particular applications or industries. Some specific tools, methods, and criteria are provided as examples. The information cited in the examples does not necessarily constitute the appropriate or only information pertinent to a specific assessment.

References

1. J. Watts, J. Hall, Chapter 72 Introduction to fire risk analysis, in *SFPE Handbook of Fire Protection Engineering*, 5th edn., (SFPE, Gaithersburg, 2016)
2. SFPE, *Engineering Guide to Performance-Based Fire Safety Design*, 2nd edn. (Quincy, National Fire Protection Association, 2007)
3. SFPE, *Engineering Guide to Fire Risk Assessment* (SFPE, Bethesda, 2006)
4. NFPA, *NFPA 551, Guide for the Evaluation of Fire Risk Assessments* (NFPA, Quincy, 2022)
5. ASME, *RA-S: 2008: Level 1/Large Early Release Frequency Probabilistic Risk Assessment for Nuclear Power Plant Applications ATIONS* (ASME, New York, 2008)
6. ISO, *ISO 16732-1:2012: Fire Safety Engineering—Fire Risk Assessment* (ISO, Geneva, 2012)

Chapter 2
Risk, Fire Risk, and Fire Risk Assessment

Risk is the potential for realization of unwanted adverse conditions, considering scenarios and their associated likelihoods and consequences. Specifically, fire risk can be defined as a quantitative or qualitative measure of fire incident loss potential for fire protection engineering applications in event likelihood and aggregate consequences.

Fire risk assessment is the process of estimating and evaluating risks associated with fires affecting buildings, facilities, or processes. The method includes evaluating relevant fire scenarios with associated frequencies and consequences using one or more acceptance criteria. In practice, fire risk assessment is used for:

- Selecting an appropriate design considering the fire risk and cost associated with various alternatives
- Managing the fire risk in a building, facility, or process
- Informing resolutions of a regulatory process, such as evaluating the risk associated with code compliance, determining acceptable configurations in risk-informed/performance-based applications.

There can be several ways of assessing fire risk. Therefore, the way risk is defined in an application is based on the study's specific objectives. For example:

- If the objective is life safety and there is concern about human fatalities in a building, risk could be measured in terms of the potential number of deaths per year.
- If the objective centers on property protection, the risk should be measured based on the potential financial value of losses per year.

In general terms, the risk parameter is measured in "outcomes per unit of activity," where the "outcome" is the potential number of unwanted events (e.g., number of fatalities), and the unit of activity is often a measure of time (e.g., a year).

SFPE Guide to Fire Risk Assessment, The Society of Fire Protection Engineers Series, https://doi.org/10.1007/978-3-031-17700-2_2

2.1 The Concept of the Risk

The risk from a particular hazard associated with a building, facility, or process may be explained as the entire domain of potential scenarios. Each scenario is represented as:

- A description of the scenario (S_i).
- An estimation of the frequency (λ_i), which refers to the characterization of how often the scenario is expected to occur. This likelihood is often represented with a frequency or a probability depending on the application.
- Characterization of the consequences (C_i).

This combination of variables (λ_i and C_i) fully characterize each scenario (S_i). The total risk for the facility can then be determined by the sum of the risk of all scenarios [1]. Consistent with the definitions above, the total risk can be expressed quantitatively, as shown in Eq. 2.1.

$$Risk = \sum_{All\,S_i} \lambda_i \cdot C_i \tag{2.1}$$

where:

λ_I = frequency of the *i*th scenario
C_i = consequence of the *i*th scenario.

Sections 2.3 and 2.4 provide a practical interpretation of these parameters. Knowing the complete set of parameters, the risk may be expressed quantitatively in several ways. The simplest form of a quantitative risk expression is the sum of the products for each scenario according to Eq. 2.1. The benefit of expressing risk in this way is that hazards are easily comparable and evaluated using a risk matrix (described in Sect. 7.6). At the same time, this method of expressing the risk does not fully represent the nature of the risk for the facility. For example, it does not provide information about the magnitude of the consequences for each scenario concerning the individual frequency. In other words, without a description of risk, insights generated by the analysis supplementing the results, a resulting risk value will not distinguish between high likelihood/low consequence events and low likelihood/high consequence events. Risk can also be expressed semiquantitatively or qualitatively by characterizing the frequency and consequences to be compared and ranked.

2.2 Scenarios

For fire risk assessment purposes, a scenario is a term used to describe a series of events that may lead to an undesired consequence per the objectives used for the basis of the assessment. The set of elements characterizing a scenario often includes fire initiation, propagation, and mitigating fire safety features available in the building, facility, or process under evaluation. It may also include elements related to human behavior and evacuation. The scenario considers the fire causing the threat and the exposed items protected for a specific set of possible events. These elements typically include the frequency of occurrence of each scenario, the hazards associated with each scenario, and the mitigating fire safety features to provide prevention or protection against fire and the potential consequences of the event. The effect of mitigating fire protection features is typically considered by applying a conditional probability to the base frequency and modifying the expected consequences. The number of scenarios can often be extensive.

Scenarios can be grouped in "clusters." This is sometimes necessary to support meaningful assessment of frequency and consequences and permit the universe of possible fires in a building to be grouped into manageable scenarios to be included in the assessment [2].

The term "fire scenario" describes how the fire develops, including ignition, growth, and extinguishment. It also considers the context, for example, the fire location and other factors needed to describe the exposure to the items to be protected for estimating the consequences (C_i) (see Sect. 3.6).

2.3 Frequency

In the context of this guide, frequency captures the likelihood of a fire occurring as the number of events that occur within a specific time interval. Frequency is the ratio of the number of times an event occurs in a time period (e.g., the number of fires per year).

The frequency can be further characterized in the analysis with a set of applicable conditional probabilities (P_i'') for each scenario (i). As such, the frequency can be expressed as follows:

- Initiating event frequency (λ_{init}) is the frequency of ignition for scenario i.
- The scenario frequency can be expressed as ($\lambda_i = \lambda_{init} \cdot P_i$) where the conditional probability represents events affecting the fire scenario progression, often associated with fire growth, detection, and suppression activities.

Conditional probabilities are necessary to describe how events may result in potential consequences. At the same time, conditional probabilities are not always required. For example, the tolerable or acceptable risk levels associated with scenarios with relatively low frequencies or consequences may not need to be further refined with conditional probabilities (i.e., conditional probabilities are assumed to be a value of 1.0).

In addition to the frequency and consequences of the postulated fire scenario, conditional probabilities are used in the context of fire risk to account for factors explicitly included in the quantification process. For example, a conditional probability may be included to characterize the failure of a suppression system given a fire, the variation of egress outcomes, etc. As such, these parameter(s) are multipliers to the initiating event frequency characterizing each scenario. This concept is further discussed later in Chaps. 10 and 11.

2.4 Consequences

In a fire risk assessment, the term "consequences" involves determining the potential impacts of a fire scenario. Depending on the objective of the risk assessment, the consequences may be expressed differently (i.e., using different units). Examples of consequences used in fire risk assessments include monetary loss per accident, numbers of injuries or fatalities, damaged building floor area, and business downtime.

2.5 Interpretation of the Risk Parameters

Consider a typical scenario consisting of the following progression of events: ignition, fire growth/propagation, detection, suppression, and resulting consequences. Each of these elements can be associated with a parameter in the risk equation described earlier as follows:

- Ignition can be quantitatively captured in the frequency term (e.g., a fire ignition frequency). This can be expressed in terms of the "number of ignitions" per unit time.
- Fire growth or propagation can be expressed in terms of a conditional probability (i.e., the probability of fire growth given ignition). This may be modified by compartmentalization strategies or limitations on available fuel load (e.g., limiting the use of combustible building materials).
- Detection and suppression can also be represented with a conditional probability (i.e., probability of detection or suppression at a point in time given fire ignition and growth).
- Given the use of conditional probabilities, an event starting with ignition may have several outcomes (i.e., different consequences). For example, successful suppression will result in a specific consequence. On the other hand, failure of suppression will result in a different set of consequences.

The conceptual example was limited to detection and suppression features. The concept of conditional probabilities can be expanded to capture most of the critical elements in the fire safety program for a facility, including fire prevention, passive fire protection, egress strategies, etc. These conditional probabilities have the practical effect of reducing the frequency of ignition events that may eventually progress to higher consequences, resulting in a lower fire risk depending on the effectiveness of the fire safety features included.

In semiquantitative or qualitative assessments, each term in the risk equation is similarly interpreted. The frequency refers to the likelihood of fire initiation. The conditional probabilities capture the scope and effectiveness of a fire safety program (i.e., mitigative strategies). The consequences refer to the expected damage or losses generated by the fire. As such, the risk equation provides a comprehensive framework for treating quantitatively, semiquantitatively, or qualitatively all elements of a fire scenario.

References

1. S. Kaplan, B. Garrick, On the quantitative definition of risk. Risk Anal. **1**(1), 11–27 (1981)
2. ISO, *ISO 16733-1:2015: Fire Safety Engineering—Selection of Design Fire Scenarios and Design Fires—Part 1: Selection of Design Fire Scenarios* (ISO, Geneva, 2015)

Chapter 3
Overview of the Fire Risk Assessment Process

This chapter provides a general overview of a fire risk assessment process. The flowchart presented in Fig. 3.1 is a summary of this process. For clarity, not all the possible interactions between activities are represented with arrows in the flowchart. It is expected that a fire risk assessment may have multiple interactions and iterations between tasks that analysts will have to manage as the project progresses and is applied throughout the life of the facility or process for which it is developed.

The process depicted in Fig. 3.1 is intended to cover four distinct phases of a fire risk assessment:

- *Phase 1 – Planning (Chaps. 4, 5, 6, and 7):* The first four activities are associated with the planning phase of a fire risk assessment. These activities are intended to define the scope and objectives clearly, collect the information necessary to perform the analysis, identify the risk assessment methods to be used, and define the acceptance or tolerance criteria governing the process.
- *Phase 2 – Execution (Chaps. 8, 9, 10, 11, 12, and 13):* Following the planning phase, the risk assessment proceeds with the technical work, including a hazards analysis, the definition, characterization of the scenarios, and the risk evaluation. This part of the execution phase is referred to in this guide as "risk assessment," that is, the systematic use of information to identify sources and estimate the risk. This phase is identified with a gray box in Fig. 3.1. This process is iterative as the analysis is expected to identify scenarios in which analytical refinements or design improvements/physical modifications are necessary to reduce risk. Such conditions will unavoidably require a reevaluation of the risk after incorporating the analytical refinements or physical changes.
- *Phase 3 – Risk Communication (Chap. 14):* Once the risk evaluation process is completed, the next phase in the process is risk communication.
- *Phase 4 – In-Service (Chap. 15):* The fourth and final phase is residual risk management and monitoring. In this phase, the assumptions and conditions governing the risk are identified and monitored throughout the facility's operational life to identify configurations associated with risk increases that may not be mitigated.

SFPE Guide to Fire Risk Assessment, The Society of Fire Protection Engineers Series, https://doi.org/10.1007/978-3-031-17700-2_3

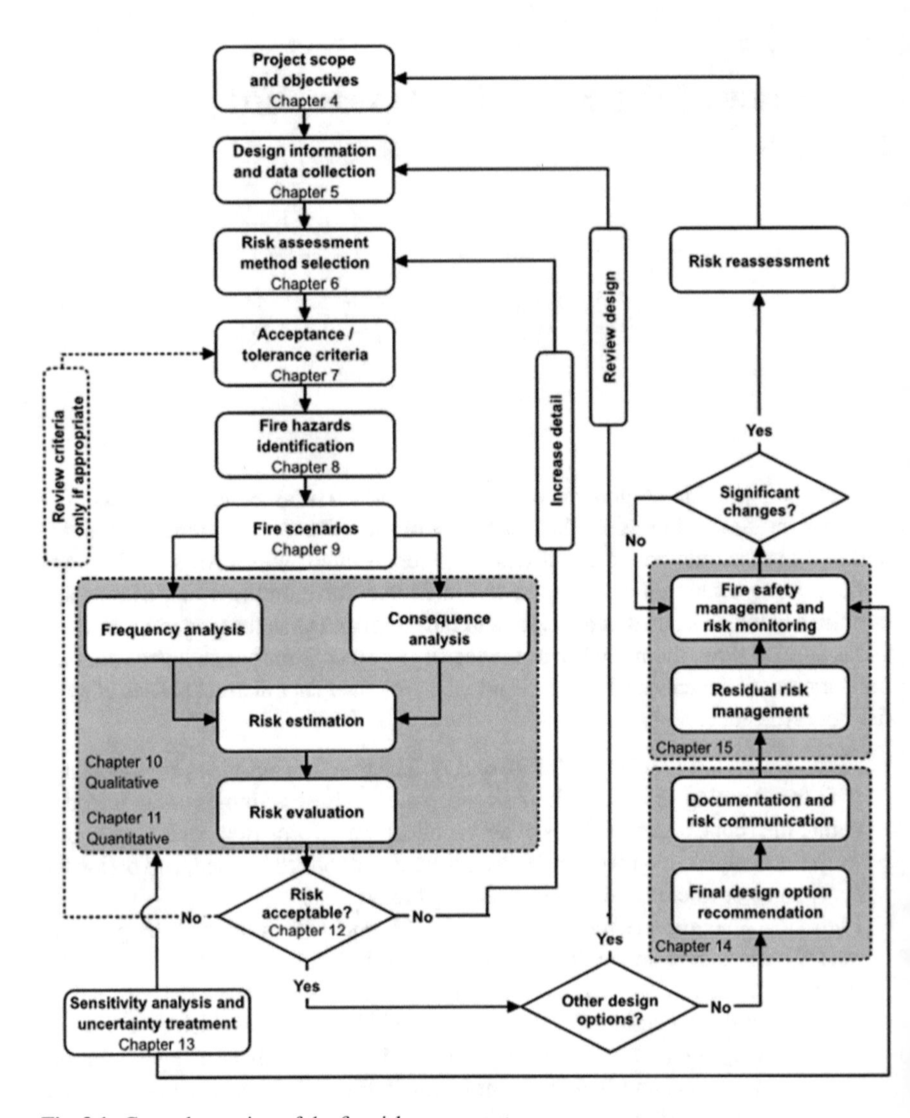

Fig. 3.1 General overview of the fire risk assessment process

Each activity in Fig. 3.1 is briefly introduced in this section and described later in this guide.

3.1 Project Scope and Objectives

The first step in the planning phase of the fire risk assessment process is to define the project scope and objectives. The project scope refers to the physical boundaries, hazards expected to be captured by the assessment, and the different stakeholders involved in the decision-making process.

Project objectives refer to the specific goals that need to be achieved (e.g., life safety, property protection). In practice, this is defined by clearly stating the purpose of the analysis.

3.2 Design Information and Data Collection

The second step in the planning phase is the project information step, which involves collecting information that will form the basis of the fire risk assessment. For example, this includes building or facility design and layout drawings, the design, and condition of any fire safety system(s), type of fire hazards, type of occupancy, and any regulatory framework applicable for conducting the study.

3.3 Risk Assessment Method Selection

The third step in the planning phase is risk assessment method selection. Risk can be assessed using qualitative, semiquantitative, or quantitative methods. Some industries may have documented methodologies that are agreed upon between the facility owner/operators and the authority having jurisdiction (AHJ). In such cases, the method selection may be limited to the technical approach to complete various tasks within a predefined methodology.

3.4 Acceptance or Tolerance Criteria

The fourth step in the planning phase is the acceptance/tolerance criteria step. The term "acceptance" refers to thresholds recognized by the AHJ defining risk levels that should be met for operation. On the other hand, the term "tolerance" refers to the willingness of facility owners and operators to assume certain risk levels. Risk values calculated for a specific application are, in most practical applications, compared against a criterion that establishes acceptable risk levels. This risk criterion should be communicated and accepted by the AHJ, so the risk assessment process has a clear framework to support risk-informed decisions made and agreed upon by the stakeholders. The acceptance or tolerance criteria should be in alignment with applicable codes and standards or regulatory requirements.

3.5 Fire Hazards Identification

The first step in the execution phase is the fire hazards identification step. A hazard is a condition or physical situation with a potential for harm. In this specific application, fire hazards (or other related conditions such as explosions) refer to induced damage, including injuries, loss of life, property losses, business interruptions, etc. This step sets the context for the remainder of the scenario description.

3.6 Fire Scenarios

The second step in the execution phase refers to the process of selecting and describing fire scenarios. A fire scenario is a set of elements characterizing a fire event. The elements are ultimately reflected in the frequency and consequence terms. The list of possible scenarios should be developed so that the range of conceivable hazards and consequences is represented adequately by a concise number of scenarios.

3.7 Frequency Analysis

The third step in the execution phase is frequency analysis which characterizes the likelihood of each fire scenario within the scope of the fire risk assessment, that is, how often the selected scenarios may occur. In practice, this is expressed in a frequency (i.e., number of events per unit of time) or probability in quantitative applications. Frequencies can also be assessed qualitatively, for example, a high/medium/low probability of an event. The fire scenarios are eventually classified in categories consistent with the risk assessment's acceptance criteria regardless of the frequency characterization.

3.8 Consequence Analysis

The fourth step in the execution phase is consequence analysis. From the perspective of a fire risk assessment, the term "consequence" refers to the outcome of the fire event. Therefore, consequence analysis involves determining the potential impact of a fire scenario directly affecting the study's objectives. To estimate the consequences, it may be necessary to break down other events to describe possible outcomes on the condition of the fire scenario. Quantitatively, consequences can be expressed in loss (e.g., financial/monetary loss, number of fatalities). Consequences can also be assessed qualitatively. As in the case of frequencies, consequences are also expressed in categories consistent with the risk assessment criteria.

3.9 Risk Estimation

The fifth step in the execution phase is risk estimation, which refers to assessing the risk contribution of each scenario by considering the combination of frequencies and consequences to present this information in an appropriate format. In a quantitative risk assessment, the resulting risk values are calculated numbers generated by a mathematical model that combines frequencies and consequences. Risk estimation may be evaluated against qualitatively expressed goals governing combinations of frequency and consequences in a qualitative risk assessment.

3.10 Sensitivity Analysis and Uncertainty Treatment

The sixth step in the execution phase is uncertainty treatment. A fire risk assessment will often require simplifying assumptions, limited data availability, engineering judgment, and analytical and empirical models. These, among other factors, will unavoidably introduce uncertainty into the resulting risk values. Therefore, a comprehensive step focusing on uncertainty treatment is necessary to identify these areas and assess the impact of these uncertainties on the assessment results. Of interest is identifying areas where final decisions, conclusions, and recommendations may be affected by the uncertainties identified in the analysis.

3.11 Risk Evaluation

The seventh step in the execution phase is risk evaluation. Multiple design options involving various combinations of risk mitigation measures may equate to a level of risk that is tolerable or acceptable to the stakeholders. However, there will be situations in which the risk estimation process suggests that design changes, further analysis, or physical modifications may be necessary to reduce risk. Incorporation of these solutions in the analysis will require reestimation and evaluation of risk results. The risk evaluation process compares the risk values associated with various options and provides the information necessary for supporting the decision-making process. The risk evaluation should also consider the results from the uncertainty and sensitivity analysis.

Risk evaluation refers to the process of evaluating a resulting risk level for determining if it meets the thresholds established as part of the acceptability criteria. This determination is made by comparing the resulting risk with the predetermined acceptance criteria and evaluating the factors (e.g., design options) influencing the consequent risk.

3.12 Risk Acceptance

The eighth step in the evaluation process is risk acceptance, which refers to evaluating a resulting risk level to determine if it meets the thresholds established as part of the acceptance/tolerability criteria. An acceptable risk refers to a low enough value that the fire scenario does not warrant further study to implement different fire protection alternatives to lower the risk or remove any conservatisms included in the analysis. This determination is made by comparing the resulting risk with the predetermined acceptance criteria and evaluating the factors (e.g., design options) influencing the resulting risk.

3.13 Final Design Option Selection

The ninth and final step in the execution phase is the final design option selection. The final design(s) or decision can be selected as the option that delivers an acceptable level of risk at the lowest cost. Alternatively, the chosen option provides the greatest risk reduction at a given budget if the acceptability threshold is met. In addition, robustness (i.e., reliance on any single risk mitigation measure) should be considered in the final design option selection process. The final decisions should also consider the results from the uncertainty treatment.

3.14 Documentation and Risk Communication

The third phase of the fire risk assessment process is the risk communication phase. Following the selection of the final design option, analysis, conclusions, and insights from the fire risk assessment should be properly documented and communicated to the stakeholders in a format and to the extent that is compatible with their needs and requirements. It is necessary to communicate the relevant details regarding the scope, method, limitations, and conclusions so that the stakeholders can take all actions required to implement and manage the design option selected and keep the risk mitigation measures effective throughout the life cycle of the building or process. For example, the fire risk assessment results may be heavily influenced by certain active fire protection features. Performance and reliability assumptions of these features may include risk monitoring activities such as specific ongoing inspection and maintenance requirements that should be communicated.

3.15 Residual Risk Management

The first step in the maintenance phase is residual risk management. Residual risk refers to a remaining level of risk that has not been removed but is still acceptable or tolerable to stakeholders. This residual risk needs to be properly documented and managed to ensure that the levels remain acceptable over time. A tolerable or acceptable risk level might be retained but should not be ignored as its acceptability or tolerability may change over time. The affected party may decide to transfer the residual risk (fully or partially) to a third party, usually an insurer, should they deem it too significant from business continuity, property protection, or other perspectives.

3.16 Fire Safety Management and Risk Monitoring

The second step in the maintenance phase is risk monitoring. The management of partially or fully retained residual risk includes risk monitoring activities such as fire prevention, inspection, and maintenance regimes, staff training, drills that have been accounted for in the risk assessment. These activities aim to protect the original assumptions and maintain the validity of the fire risk assessment. If significant changes are identified, their impact on the fire risk should be analyzed, and its acceptability/tolerability reevaluated.

Chapter 4
Project Scope and Objectives

This chapter provides guidance on factors influencing the identification of the scope and objectives for the fire risk assessment process. The project scope and objectives provide the foundation for the insights derived and the breadth and depth of the assessment. As a minimum, the risk assessment should be designed to identify key risk drivers and the appropriate fire protection strategies and mitigations.

4.1 Scope

The project scope refers to the assessment boundaries and the different stakeholders involved in the decision-making process. The scope of a fire risk assessment is often defined based on one or more of the following elements:

- Description of the physical location and boundaries of the building, facility, or process
- Types of fire hazards considered in the assessment
- Elements of the fire protection system under evaluation
- Occupancy characteristics
- Stakeholders and external parties
- Regulatory requirements and any other external factors

4.1.1 Physical Location and Boundaries

Physical location and boundaries are often the first elements governing the scope of a risk assessment. Physical boundaries limit the space, both for the design and where the design will be implemented. Examples of physical boundaries include the property lines of a site or complex, the exterior walls of a building or vehicle, and

SFPE Guide to Fire Risk Assessment, The Society of Fire Protection Engineers Series, https://doi.org/10.1007/978-3-031-17700-2_4

the internal boundaries of the part of a building where an activity or process is conducted or where a renovation project is to focus.

The scope can also be defined in terms of temporal boundaries. Temporal boundaries are limits in time rather than space. For example, when a property is open and operating entails different risks than when it is closed.

4.1.2 Hazards

A hazard may relate to a physical object, a process, a person(s), a physical configuration, or an environmental condition. In this guide, the term "fire hazard" is the primary hazard whose potential for harm arises from unwanted fire. Examples include ignition sources igniting combustible materials that may propagate a fire, explosion sources, etc. An early step in a fire risk assessment consists of comprehensive hazard identification intended to ensure that all relevant hazards are captured. This ensures that the risk contribution of each of the hazards is assessed. In this context, the analyst is not expected to identify all potential hazards in defining the scope of the assessment, as the identification process is part of the process. This will ensure the completeness of the assessment in terms of identifying risk contributors. Instead, the scope definition refers to identifying and justifying the precursors of such hazards that will be included in the assessment.

4.1.3 Fire Protection

Fire protection refers to installed features, alternative fire safety solutions, or strategies considered in the fire risk assessment. Typically, the risk assessment is used to characterize these features or strategies' ability to mitigate fire risk.

4.1.4 Occupancy

Occupancy refers to the occupants associated with the building, facility, or process which may potentially be physically affected by an unwanted fire event within the scope of the risk assessment. This element can also include occupants of surrounding buildings, facilities, processes, or residents of larger municipalities/cities/towns.

4.1.5 Stakeholders

Stakeholders refer to parties who have a stake in how the fire risk associated with the building, facility, or process is dealt with and involved in any consultation relating to the risk assessment process and who may be involved in any decision-making.

4.1.6 Regulatory Requirements

Regulatory requirements refer to the regulations, codes, standards, and other external factors that apply to the building, facility, or process under consideration for the fire risk assessment.

4.2 Objectives

The fire risk assessment process starts with establishing the objectives such that all stakeholders understand the purpose of the analysis. Typical objectives may include:

- Evaluating the risk levels of conforming to code or insurance requirements in support of fire safety goals such as life safety, property protection, minimizing business interruption, preservation of cultural heritage, minimizing environmental impact.
- Evaluating risk-informed strategies for achieving equivalent risk levels as conforming to code or insurance requirements when direct compliance is not possible or determined to be too costly.
- Selecting a cost-effective fire protection strategy.
- Assessing and communicating the risk associated with a specific activity, facility, system, or process in support of routine operations.

Once established, the objectives need to be translated into a risk metric agreed upon by stakeholders that can be compared with acceptable thresholds. For example, suppose the objective is to "minimize business interruption." In that case, the analyst will need to translate "business interruption" into a risk metric that can be used in the decision-making process. For example, the risk metric could be expressed in terms of "money-per-fire event" or "money-per-nonoperational day."

4.3 Example: Project Scope and Objectives Example

Section A.1 provides a conceptual example related to project scope and objectives.

Chapter 5
Design Information and Data Collection

This section describes the collection of information that will form the basis of the fire risk assessment. The information typically includes facility design layout, operation and maintenance processes, characteristics of materials within the facility, training of staff and occupants (as it relates to human response to a fire event), and fire protection systems and features. Each of these elements affects how the fire scenarios are developed and characterized. This section also describes the various information sources typically used when performing a fire risk assessment and how each source of information relates to the fire risk assessment.

5.1 Facility Documentation and Drawings: Non-fire Protection

Engineering and architectural documentation and drawings provide essential design information about the construction and operational characteristics of the building, facility, or process. From the perspective of fire risk assessment, the design information contained in documentation and drawings can also limit what types of hazards could be allowed within a building, facility, or process. This information is important to understand the arrangement, operation, and maintenance of the facility's equipment, physical barriers, and conditions. These documents and drawings include, but are not limited to:

- System specifications
- Product and system technical literature
- Operating and maintenance manuals
- Layout drawings, including structural and architectural drawings depicting relevant building features such as means of egress
- HVAC drawings, drawing depicting mechanical ventilation operational details, and the location of vents and ducts

SFPE Guide to Fire Risk Assessment, The Society of Fire Protection Engineers Series, https://doi.org/10.1007/978-3-031-17700-2_5

- Electrical drawings depicting the location of power supplies, electrical equipment, and cabling layouts
- Mechanical drawings depicting the location of piping, mechanical equipment, etc.

Engineering and architectural documentation and drawings assist in identifying the scope of the analysis and the physical configuration of the facility. In addition, these documents and drawings also provide information that may influence the hazard assessment. Examples include the location of ceiling beams identified in structural drawings that may impede a ceiling jet's flow and delay automatic smoke detection and forced ventilation flows (identified in HVAC drawings) that may affect the development of hot upper layer temperatures.

In addition to any complementary reports, models and calculations should be reviewed as appropriate. This reporting may include information about fuel type, fuel load, configuration, occupancy type, occupant density, mobility limitations, etc.

5.2 Regulatory Requirements

Regulatory requirements refer to code and standard requirements governing the design, operation, and maintenance of the building, facility, or process under evaluation. Such requirements are captured in the risk assessment to appropriately reflect their impact on the fire risk assessment and support the effective communication of the risk results to stakeholders. Regulatory requirements may also include the risk acceptability thresholds governing the risk assessment process.

5.3 Occupancy Information

Occupancy information may include personnel present in the facility continuously (e.g., employees), visitors, customers, and others in a public occupancy. Occupancy information is often used to inform the development of fire scenarios. For example, hours of operation may inform the frequency of postulated fires. Means of egress or limitations on the mobility of occupants may influence the assessment of the consequence terms. Training and procedures may inform conditional probabilities associated with human detection and suppression of fires. Examples of occupant information that may be applicable to a fire risk assessment include:

- Egress and mobility factors
- Personnel training and other applicable procedures (e.g., pre-job briefings to visitors)
- Number of occupants
- Hours of operation

- Characteristics of visitors, customers, and other expected occupants in a public facility (e.g., degree of familiarity with the facility)
- Human behavior factors
- Types of activities performed in the facility at different times

5.4 Process Documentation and Drawings

Process documentation and drawings may provide input on the frequency and consequence of a hazard's location, storage, and use. Examples include:

- Operational processes
- Piping and instrumentation diagrams (P&IDs)
- Information about hazardous materials
- Equipment information
- Drawings of equipment

5.5 Fire Protection Documentation and Drawings

Fire protection documents and drawings identify various aspects of the fire protection systems and features, including what type of detection and suppression devices are available, what type of passive fire protection systems and features are present, and their corresponding locations. The location of these devices, systems, and features may be an important scenario-specific element for consideration in a fire risk assessment.

Fire protection systems and features can be divided into two groups:

- Controls that reduce the frequency of fires by preventing or reducing the occurrence of an event.
- Systems that mitigate the damage caused by fire (i.e., those systems that are barriers to harm after the fire has ignited).

The following resources will contain information on the controls that aim to prevent or reduce the frequency of a fire:

- Fire prevention plans
- Procedures for hazardous material storage and disposal
- Fire watch processes

Fire protection systems that mitigate consequences range from fixed detection and suppression systems to human response once a fire occurs. The response of these features is a vital aspect of developing a risk assessment. Information on the effectiveness, reliability, and availability of these features is necessary to include the features' effects in the analysis appropriately.

Examples of information on fire protection features relevant to a fire risk assessment include:

- Fire protection drawings
  - Location and fire resistance of fire barriers and structural fire protection
  - Location of egress routes and exits
  - Fire alarm (detection)
  - System layout drawings
  - Suppression system layout drawings and supporting calculations

    Hydraulic calculations
    Clean agent suppression system calculations

  - Smoke control systems
- Inspection, testing, and maintenance protocols and reports
  - Automatic systems, regardless of system design, are reliable only when adequately inspected and maintained. These protocols and reports ensure that the systems reliably detect and confine fires that occur.
- Pre-fire plans and emergency response plans
  - Important to assess preparedness to fire events, especially the potential scenarios defined as part of the risk assessment
- Fire brigade and fire department information
  - Important for estimating response times and capabilities

5.6 Fire Hazard Information

Fire hazard information is necessary to support the development of fire scenarios in the fire risk assessment. Some facilities may have already completed a fire hazards analysis as part of their fire protection program. Chapter 8 provides additional guidance for identifying and characterizing fire hazards for a fire risk assessment.

5.7 Example: Design Specification and Data Collection

Section A.2 provides a conceptual example of design specification and data collection.

Chapter 6
Risk Assessment Method Selection

The selection of the risk assessment method(s) relates to the level of detail to which each scenario is described and quantified concerning the level of potential risk. In general terms, the analysis (i.e., the activities within the assessment where risk is evaluated) can range from qualitative to quantitative, including semiquantitative approaches. This is governed primarily by the level of perceived risk, which may change as the overall assessment progresses, and by regulatory bodies. In practice, the type of risk-based evaluation and level of detail should depend on the complexity of the risk and the decision-maker's needs [1]. When selecting the type of analysis, it is necessary to consider several factors, including the information available, the complexity of the facility or process under analysis, the potential deviations from code requirements and best practices, and the level of detail necessary to make a substantiated decision about the tolerability of fire risk(s).

Qualitative analysis refers to the evaluation of risk without explicit numerical quantification. In a qualitative assessment, fire risk is evaluated based on the merits of specific designs versus the postulated potential fire events. Qualitative risk methods may be appropriate for evaluating well-understood conditions associated with simple systems or configurations with established risk levels.

A semiquantitative analysis refers to the evaluation of risk with simplified quantitative elements supporting assessment. This approach may be appropriate for evaluating configurations with minor deviations from code requirements or best practices and risk trade-off implications.

Finally, a quantitative analysis is a complete explicit quantification of frequencies and consequences to produce numerical risk levels. The need for a quantitative assessment often arises when evaluating novel, challenging, or complex configurations with significant risk trade-offs. Additional factors influencing the need for a quantitative analysis include identifying significant uncertainties that need to be rigorously modeled and strong stakeholders' views and perceptions of potential risks. Alternatively, the relevant regulations may mandate a quantitative approach.

Figure 6.1 depicts an iterative approach in which the level of detail increases as the level of quantification increases. Notice that the level of effort increases as more

SFPE Guide to Fire Risk Assessment, The Society of Fire Protection Engineers Series, https://doi.org/10.1007/978-3-031-17700-2_6

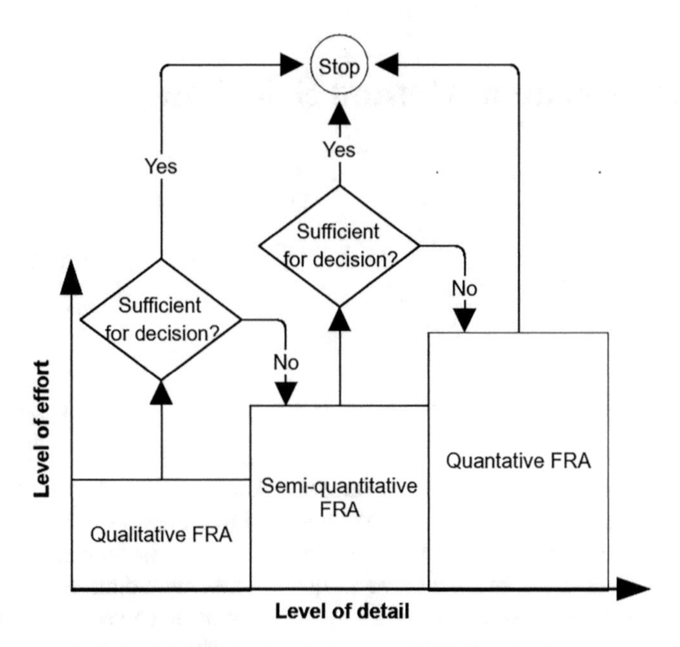

Fig. 6.1 Iterative selection of the Fire Risk Assessment (FRA) approach

quantification is required to support the conclusions. Also, a risk assessment may be developed with a combination of qualitative and quantitative approaches (i.e., semi-quantitative approach) while maintaining the rigor and analysis necessary to reach conclusions.

For more information, see Chap. 10 (Qualitative Fire Risk Estimation) and Chap. 11 (Quantitative Fire Risk Estimation).

Reference

1. SFPE, *Handbook of Fire Protection Engineering*, 5th edn. (SFPE, Gaithersburg, 2016)

Chapter 7
Acceptance or Tolerance Criteria

Fire risk assessment involves the need to establish a target risk (i.e., a criterion for tolerance or acceptability). The target risk should provide a socially acceptable level of outcome considering stakeholders' perspectives. When considering stakeholders' views, different types of risk are perceived differently, and society prefers varying levels of outcome depending on the risk characteristics. For example, potentially catastrophic risks are perceived differently than less severe ones. An occupancy where hundreds of people are at risk due to a single fire is perceived differently than an occupancy where only one person is at risk due to a single fire. Society prefers differing levels of safety for the two different occupancies. Based partly on perceptions, this preference is reflected deterministically in many global codes and regulations.

7.1 Stakeholders

Each stakeholder group may contribute to the project decision-making process. Stakeholders usually include:

- People who may experience consequences associated with the risks (e.g., harmed by fire, have a property that is damaged by fire, have businesses or jobs that are interrupted or lost due to fire) are stakeholders. Frequently, the interests of people whose only stake in a project are their potential vulnerability to harm will be represented by an AHJ.
- The facility owner is typically concerned with the full range of risk (life safety, property protection, continuity of operations, environment), with the emphasis varying by facility use (e.g., public assembly versus storage), size, and location. By necessity, the owner will have a significant focus on costs, including ongoing costs and opportunity costs. The owner may designate a building manager or a risk manager as an agent to represent their interests.

SFPE Guide to Fire Risk Assessment, The Society of Fire Protection Engineers Series, https://doi.org/10.1007/978-3-031-17700-2_7

- The occupants of a facility typically have the perspective of personal safety, that is, they are concerned that they are in a building that will provide them with a reasonable assurance of not being injured due to a fire.
- The neighbors' perspective on a facility is the concern that an event at the neighboring facility does not have a detrimental effect on themselves and their facility. The effects could be from fire, combustion products, collapse, etc.
- Federal, regional, and local governments are formed to provide for the protection of their populations. The harm need not be direct. For example, unemployment and subsequent loss of a municipality's tax base can be significant. The corresponding harm caused by excessive requirements or lack of protection services can include employers moving to more supportive areas and residents leaving because of excessive taxes or absence of other services.
- Regulators are usually employees from different levels of government, such as national and local, but their perspective is not the same as the government entity. Typically, the regulator focuses on one aspect of the risk (e.g., fire hazards) because they enforce specific regulations. Non-fire risks, which the municipality might address, are not a direct concern. Instead, the regulator may be concerned with the risk of being nonconservative. This concern is possible when a loss occurs in a facility that the regulator approved and did not explicitly meet the regulations.
- First responders (firefighters) expect that hazardous conditions may exist in a facility fire; however, they typically expect the structure to remain reasonably stable early in the fire to allow for evacuation and firefighting operations to take place.
- The insurer's primary objective is to provide risk-sharing for the building owner and tenants. Property and casualty insurance companies have different perspectives. Property insurers are primarily concerned about property and business continuity, and casualty insurers are primarily concerned about life safety.
- The designer is concerned with meeting the objectives of providing a facility that meets the requirements of the various stakeholders. The designer will typically be directed by the owner but will have to meet the requirements of regulators, insurers, and others. The designer is concerned with the engineering requirements and costs of the options to meet the acceptable risk.
- The risk manager will balance various costs, including insurance, deductibles, construction, against acceptable risk.

All stakeholders' perspectives should eventually be represented as a consistent threshold to allow risk decisions to be made and agreed.

7.2 Tolerance and Acceptance

The term "tolerance" refers to the ability or willingness to accept a specific risk level. It is used in this guide in the context of the stakeholder assuming a level of risk (e.g., a facility operator, an insurer). On the other hand, the term "acceptance" is used in this guide from the perspective of an AHJ approving a specific operational condition. Therefore, risk acceptance is the process of reviewing risk levels that are either estimated or calculated against the risk criteria established. Suppose the assessed risk is lower than the criteria established at the outset. In that case, the analysis may be considered complete. No further study of additional or alternative fire protection alternatives may be necessary. Also, as part of this process, and to further bolster risk acceptance, any conservatism in the analyses can be identified, assessed, and documented for presentation in the risk communication and monitoring stages.

7.3 Establishing Risk Criteria

A single entity should not establish risk criteria (e.g., acceptable risk thresholds) without consultation with stakeholders. This is because the acceptable risk is both a technical concept and a value judgment. All stakeholders should be included in establishing agreed-upon criteria by setting the acceptable level of risk and its metrics. This process should also be consistent with adopting a set of deterministic expectations, often expressed in the form of regulatory or code requirements that should be met to form the basis for the level of fire safety with which facilities routinely operate. That is, acceptable risk levels should represent the level of fire safety expected in facilities that meet the applicable regulatory or code requirements.

To begin establishing risk criteria, understanding the concept of "de minimis risk" is required. De minimis risk is based on the premise that there is some level of risk below which one does not need to be concerned. The idea is that stakeholders can agree upon a de minimis threshold and agree that no mitigation is required to lower the risk below that value. There are often difficulties in gaining such agreement. For example, if a proposed de minimis threshold were framed in terms of an acceptable death toll, it would generally be challenging to obtain broad agreement even over a very long period. Where there is discomfort with a proposed de minimis threshold, that discomfort may take the form of extended questioning of or challenge to the procedures and assumptions used in estimating the risk that will be compared to the threshold. Challenges may be associated with risk perception, and there may be a close examination of the degree of conservatism incorporated into the estimates. For example, a de minimis threshold for an expected-value risk measure may not be accepted if the risk includes the possibility (even with a very low probability) of a large life-loss event or exceptionally large property loss event. A life loss sufficiently large as to destroy a small community or a property loss sufficiently large as to destroy a part of the insurance industry, for example, would likely be judged in very different terms.

The ALARP (As Low As Reasonably Practicable) principle provides an approach to resolving the limitations of identifying a "de minimis" risk level. ALARP [1] is common in engineering projects and is also used for fire risk assessments [2]. De minimis risk thresholds provide a level of risk sufficiently low that it can be considered negligible. In contrast, ALARP adds a region of alternatives whose risk may be acceptable when evaluated in the context of cost and level of safety. The evaluation of costs defines the difference between what is achievable and what is reasonably achievable. In fire risk analyses, an ALARP approach might be an iterative analysis, varying or adding fire safety features within acceptable limits as defined by the AHJ to the point where additional fire safety features result in significant additional costs far exceeding the marginal benefits. According to the ALARP principle, it becomes impractical at this point to add more costly features when their benefits are very low.

In practice, when a quantitative fire risk assessment is used, the expected risk is explicitly targeted. In semiquantitative fire risk analyses, such as ranking or scoring methods [3], where additional fire safety measures are attributed indexed values, a minimum total fire safety score or set of scores is established as the acceptable level of risk. The analyst would investigate various measures to reach the minimum score or scores and stop adding or changing measures once the scores are achieved.

7.4 Tolerability and Acceptability

Figure 7.1 captures both aspects in the decision-making process: tolerability and acceptability.

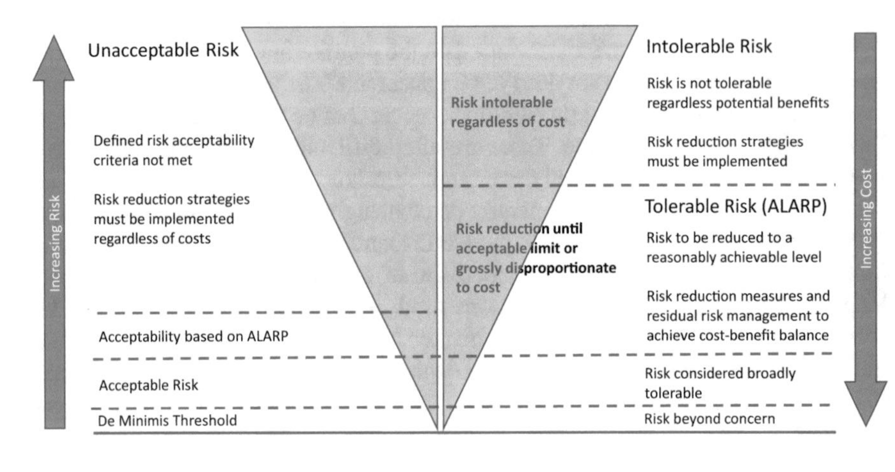

Fig. 7.1 Pictorial representation of acceptable (left triangle) and tolerable (right triangle) risk levels

Tolerability is represented on the right-hand side of the triangle and represents the perspective of all stakeholders other than the AHJ. The upper part of the tolerability triangle represents an intolerable risk for the stakeholder. This level of risk needs to be reduced. As one moves down the right-hand side of the triangle, the risk may reach a tolerable level that can be further reduced through the ALARP principle. In the ALARP region, the risk is reduced considering the cost-benefit balance until the balance reaches a limit where the cost is disproportionate to the benefit. Below that level, the risk is considered tolerable and not practical to reduce further.

The left-hand side of the triangle represents acceptability and represents the perspective of the AHJ. This is represented as a threshold value above which the risk level will not be acceptable to the AHJ. Below this threshold, the risk level is acceptable.

In some applications, there may not be a risk acceptability threshold defined by the AHJ. In such cases, when presented as a performance-based alternative, a fire risk assessment should be primarily evaluated using the ALARP principle. Therefore, the acceptable threshold is within the ALARP range of acceptable risk as defined by the AHJ. Risk is reduced as low as reasonably practicably in the ALARP region to a level agreed by the AHJ as acceptable for the application.

7.5 Risk Perception

The concept of risk perception reflects the stakeholder's needs, issues, knowledge, and values [4]. It can be a factor that influences acceptability or tolerability. For example, a high-rise hotel fire risk is perceived as potentially more catastrophic and regarded as an uncontrollable risk when compared to a single-family home. It follows that the acceptable level of fire risk in a high-rise hotel would differ from that of a single-family home. Wolski [5] described the mechanisms for risk perception and recommended an approach for quantifying it in the form of several multiplicative risk factors for adjusting the threshold. These risk factors are summarized in Table 7.1. Existing risk thresholds (i.e., those already established and agreed upon by stakeholders in a given industry) have already accounted for risk perception, and the resulting levels have been established to represent them. As a conceptual example to apply the factors in Table 7.1, a baseline risk threshold of 1.0E-3 events per year is assumed. If stakeholders perceive consequences of fire accidents to be "involuntary" (i.e., it is not expected that activities can result in a fire accident) in relation to the "Volition" risk factor, the risk conversion factor from Table 7.1 is 1/100, and the baseline risk threshold may be reduced by up to $1.0\text{E-}3 \times 1/100 = 1.0\text{E-}5$.

In many cases, several risk factors can play a role in the same occupancy simultaneously. For example, people prefer more safety when in a high-rise hotel, where they perceive not only a greater potential for the "Severity" risk factor but have less impact on the "Controllability" risk factor than in their single-family homes. Notice that the baseline risk threshold should be agreed upon first before the risk perception factors are applied. In practice, establishing a very low baseline before applying the

Table 7.1 Summary of potential risk conversion factor ranges

Risk factors	Scale	Risk conversion factors[a]	Comment
Volition (i.e., a choice or decision made)	Voluntary – involuntary	1–1/100	An involuntary activity that results in a fire is perceived as having a higher risk for which a tolerable or acceptable risk threshold level may be reduced by a factor up to 1/100
Severity	Ordinary – catastrophic	1–1/30	The potential for catastrophic consequences resulting from a fire scenario may suggest reducing the tolerable or acceptable risk threshold level by a factor up to 1/30
Effect manifestation	Immediate – delayed	1–1/30	Immediate consequences resulting from a fire scenario may suggest reducing the tolerable or acceptable risk threshold level by a factor up to 1/30
Familiarity	Common – dread	1–1/10	Activities perceived as unfamiliar may require reducing the tolerable or acceptable risk threshold level by a factor up to 1/10
Controllability	Controllable – uncontrollable	1/5–1/10	The inability to control the consequences of fire scenarios (e.g., expected scenarios that may overcome existing mitigation strategies) may suggest reducing the tolerable or acceptable risk threshold by a factor of 1/5 to 1/10.
Benefit	Clear – unclear	Risk is roughly proportional to the third power of its benefit	Stakeholders may accept a higher level of risk (i.e., resulting in a higher threshold) if there is a clear benefit from their activities
Necessity	Necessary – luxury	1	No change in baseline due to perception
Exposure pattern	Continuous – occasional	1	No change in baseline due to perception
Origin	Natural – man-made	1–1/20	The potential for consequences resulting from a "man-made" fire scenario may suggest reducing the tolerable or acceptable risk threshold level by a factor up to 1/20

[a]Range of suggested values

factors may result in unreasonably low-risk levels that may never be met. In the presented conceptual example, the baseline was set in the "remote" frequency classification in the risk matrix, which, based on risk matrices that have been already established in different industries (see Sect. 7.6), represents a transition region between acceptable/tolerable and unacceptable/intolerable risk.

Risk perception does not always need to be accounted for explicitly. For example, existing risk matrices (i.e., risk matrices already established and agreed upon by stakeholders in an industry) most likely have already accounted for risk perception.

7.6 The Risk Matrix

A common approach for representing a risk assessment is using a risk matrix. A risk matrix is also a tool often used to summarize stakeholders' viewpoints on risk levels (e.g., the combination of high consequence-low probability events). In addition, it is an effective visual tool for communicating risk and serves as a basis for decision-making in fire risk assessment.

The risk matrix has likelihood (typically frequency or sometimes probability) on one axis and consequences on the other axis. Both the frequency and consequences are classified into categories.

Frequencies and consequences governing the risk are threshold levels for decision-making purposes and are independent of the fire protection system and features governing the risk of individual scenarios. The risk of each of the scenarios included in the risk assessment is determined considering the fire protection system and applicable features. Such systems and features provide the level of safety as expressed by its corresponding risk compared to the risk levels in the matrix for tolerability and acceptability decisions.

Table 7.2 provides guidance on practical frequency levels that can be used to develop a risk matrix in both qualitative and quantitative terms. Note that the rankings provided are examples, and these ranges may be revised for specific risk assessments.

Table 7.2 Typical frequency levels in a risk matrix [6]

Ranking	Description	Frequency
Frequent	Likely to frequently occur during the lifetime of an individual item or very often in the operation of a large number of similar items	Greater than1.0/year
Probable	Will occur several times during system life or often in the operation of a large number of similar items	1.0^{-1}/year to 1.0/year
Occasional	Likely to occur sometime in the lifetime of an item or will occur several times in the operation of a large number of similar items	1.0^{-2}/year to 1.0^{-1}/year
Remote	Unlikely, but possible to occur in the lifetime of an individual item. It can be reasonably expected to occur in the operation of a large number of similar items	1.0^{-4}/year to 1.0^{-2}/year
Improbable	Very unlikely to occur. It may be possible but unlikely to occur in the operation of a large number of similar items	1.0^{-6}/year to 1.0^{-4}/year
Incredible	Events that are not expected to occur	Less than 1.0^{-6}/year

Similar to frequencies, consequences can be categorized into different levels. Typically, these levels range from negligible to catastrophic. Given that consequences can be expressed in different terms (e.g., different units) depending on the application, it is practical to represent them with normalized values expanding various orders of magnitude. This has the following practical implications:

- Consequences levels are rarely linear. They often range from negligible, consisting of minimal impact, to catastrophic consisting of very large implications. A financial/monetary value can easily replace the normalized values, the number of injuries or deaths, etc.
- Normalized consequences allow for the interpretation of risk in frequency terms. This may ease the risk evaluation and communication process. For example, setting the catastrophic consequence to 1.0 allows for decision-making based on the likelihood of event occurrence as defined by the frequency levels, which have already been defined in terms of the expectancy of experiencing an event in an individual's lifetime.

Ultimately, each consequence level is mapped to a number between a value close to 0 and 1 representing the damage caused by the fire scenario. An assessment should inform these numbers of the consequences relative to the potential number of injuries, fatalities, property damage, and business interruption losses that may occur. As an example, the consequences in Table 7.3 are expressed as a normalized value. If the risk is represented as the death or injury rate due to fire per year, the consequence term may be the number of death or injuries. Note that the rankings provided are examples, and these ranges may be revised for specific risk assessments.

Utilizing the information in Tables 7.2 and 7.3, two conceptual examples of a risk matrix are provided in Tables 7.4 and 7.5. In these examples, the header rows represent the different consequence levels. The first two columns represent the frequency categories. The resulting risk level is then obtained from the different combinations of frequency and consequences in the matrix. The matrix can be qualitative or quantitative. It is noted that quantitatively, different frequency and consequence categories are populated with numerical values. Note that the rankings provided are examples, and these ranges may be revised for specific risk assessments.

Depending on the specific application, the matrix could be interpreted as follows:

- "Negligible" or "Marginal" consequences can be accepted (or tolerated) as the "cost of doing business" as it relates to routine equipment failures and personnel accidents not requiring medical treatment beyond first aid. However, repeated "Negligible" or "Marginal" consequences may indicate insufficient or deteriorating fire strategy and management and should be addressed.
- "Improbable" and "Incredible" frequencies can also be accepted (or tolerated) as those events are highly unlikely to occur. The resulting risk values in this range consider the effects of fire protection features. The analyst should determine if there is a margin with and without these features. Fire protection features lowering risk values will need to be monitored routinely to ensure their effectiveness

Table 7.3 Typical consequence levels in a risk matrix

Ranking	Description	Normalized consequences
Negligible	The impact of loss is so minor that it would not have a discernible effect on the occupants, facility, operations, or the environment. Examples of negligible consequences may include: No recordable/reportable event (i.e., event does not result in any work-related injury or illness requiring medical treatment beyond first aid) Property losses consistent with failures address with routine budgeted maintenance activities	Less than 1.0^{-5}
Marginal	The loss has a limited impact on the facility, which may have to suspend some ancillary operations briefly. Some monetary investments may be necessary to restore the facility to full operations. Minor personal injury may be involved. The fire could cause localized reversible environmental damage. Examples of marginal consequences may include: A recordable/reportable event (i.e., event included mitigation consistent with the state of industry practice such as sprinkler activation and no injuries beyond those requiring first aid) Property losses consistent with those associated with damage limited by the effective operation of fire protection mitigation strategies	1.0^{-5} to 1.0^{-3}
Major	The loss has a significant impact on the facility, which may have to suspend main operations for a limited time. Significant monetary investments may be necessary to restore to full operations. Multiple minor personal injuries and/or a single severe injury are involved. The fire could cause significant localized but reversible environmental damage	1.0^{-3} to 1.0^{-1}
Critical	The loss has a critical impact on the facility, which may have to suspend operations for a prolonged period. Major monetary investments may be necessary to restore to full operations. Multiple severe personal injuries and/or a single fatality are involved. The fire could cause extensive but reversible environmental damage	1.0^{-1} to 1.0
Catastrophic	The loss has a high impact on the facility, which may have to suspend operations permanently. Monetary investments reaching total facility cost may be necessary to restore to full operations. Multiple deaths may be involved. The fire could cause irreversible environmental damage	1.0

over the facility's operational life to minimize periods without protection that may be associated with higher risk levels.

- Risk values higher than 1.0E-3 may not be accepted or tolerated. Under this interpretation, which is based on normalized consequences, these values suggest frequencies of occasional events (or higher), which should be addressed by improvements in the design or the fire protection strategy.

Table 7.4 Qualitative risk matrix

	Consequence				
Frequency	Negligible	Marginal	Major	Critical	Catastrophic
Frequent	Acceptable	Further evaluation	Not acceptable	Not acceptable	Not acceptable
Probable	Acceptable	Further evaluation	Not acceptable	Not acceptable	Not acceptable
Occasional	Acceptable	Acceptable	Further evaluation	Not acceptable	Not acceptable
Remote	Acceptable	Acceptable	Acceptable	Further evaluation	Further evaluation
Improbable	Acceptable	Acceptable	Acceptable	Acceptable	Further evaluation
Incredible	Acceptable	Acceptable	Acceptable	Acceptable	Acceptable

Table 7.5 Quantitative risk matrix

		Consequence				
		Negligible	Marginal	Major	Critical	Catastrophic
Frequency		1.0E-06	1.0E-04	1.0E-02	5.0E-01	1.0E+00
Frequent	1.0E+00	1.0E-06	1.0E-04	1.0E-02	5.0E-01	1.0E+00
Probable	1.0E-01	1.0E-07	1.0E-05	1.0E-03	5.0E-02	1.0E-01
Occasional	1.0E-02	1.0E-08	1.0E-06	1.0E-04	5.0E-03	1.0E-02
Remote	1.0E-04	1.0E-10	1.0E-08	1.0E-06	5.0E-05	1.0E-04
Improbable	1.0E-06	1.0E-12	1.0E-10	1.0E-08	5.0E-07	1.0E-06
Incredible	1.0E-08	1.0E-14	1.0E-12	1.0E-10	5.0E-09	1.0E-08

- Finally, fire scenarios associated with risk values in the range of 1.0E-5 to 1.0E-3 may require further evaluation. The scenarios that credit the fire protection systems to reduce the consequences may be in this regime. Therefore, the evaluation should ensure that:
  - Risk insights are appropriately obtained. This specifically refers to the factors driving the risk numbers. These factors may point to the fire protection capabilities that may need to be improved.
  - There is a margin in the risk results. Sometimes, conservatism affects the input parameters' risk values or the models used to represent the scenarios. Identifying such conservatisms suggests a margin in the analysis that can be used to justify a final decision.
  - An additional level of safety is provided if necessary. In situations where risk insights suggest low margins, additional fire protection features may be recommended.
  - Appropriate fire protection strategies are monitored to maintain acceptable risk levels.
  - If available, defense-in-depth measures address these fire scenarios (i.e., fire protection strategies beyond those explicitly included in the analysis).

For example, a facility has a plastics extruder with injection and blow molding presses and a warehouse. The loss (i.e., consequence) associated with either the manufacturing or the warehouse may be classified as major for the facility operator. However, assuming a lack of clear space and fire-rated construction between manufacturing and the warehouse, the consequence of loss may be critical or catastrophic. Therefore, depending on the frequency level, reliance on fire protection strategies (e.g., fire-rated constructions, automatic sprinklers) is necessary for maintaining tolerable risk levels.

7.7 Risk Matrix Examples

The conceptual risk matrices described in Sect. 7.6 are based on recommended frequency levels and consequences currently used in various applications. This section provides examples of applications identified as part of the research consulted when developing this guide.

- Barry [7] cites the following ranges for risk tolerance for life safety:
  - 10^{-3} to 10^{-4} per year for major injury or fatality potential for plant personnel working within the direct boundary of the facility or operation under evaluation
  - 10^{-4} to 10^{-5} per year for major injury or fatality potential for plant personnel working beyond the direct boundary of the facility or operation under evaluation
  - 10^{-5} to 10^{-6} per year for major injury or fatality potential for the general population beyond the plant or facility boundary
  - $<10^{-6}$ per year for multiple injury or fatalities potential in highly populated areas outside the boundaries of the plant or facility.
- The US military standard MIL-STD-882E [8], which is standard practice for safety in engineering systems, provides a method for identifying, classifying, and mitigating hazards. The document describes the following generic risk matrix to use as a starting point in specific applications:
  - Consequence levels:
    (a) Catastrophic: death, permanent total disability, irreversible significant environmental impact, or monetary loss equal to or exceeding $10 M.
    (b) Critical: permanent partial disability, injuries, or occupational illness that may result in hospitalization of at least three personnel, reversible significant environmental impact, or monetary loss equal to or exceeding $1 M but less than $10 M.
    (c) Marginal: injury or occupational illness resulting in one or more lost workday(s), reversible moderate environmental impact, or monetary loss equal to or exceeding $100 K but less than $1 M.
    (d) Negligible: injury or occupational illness not resulting in a lost workday, minimal environmental impact, or monetary loss less than $100 K.

– Frequency levels:

(a) Frequent: often in the life or continuously experienced in a fleet or inventory.
(b) Probable: several times in the life of an item or frequently in a fleet or inventory.
(c) Occasional: sometimes in the life of an item or several times in a fleet or inventory.
(d) Remote: unlikely but possible in the life of an item or unlikely but reasonably expected to occur in a fleet or inventory.
(e) Improbable: not expected to occur in the life of an item or unlikely but possible in fleet or inventory.
(f) Eliminated: incapable of occurring.

With these consequence and frequency levels, the qualitative risk matrix summarizing the risk presented is identified in Table 7.6.

- The US Nuclear Regulatory Commission considers and approves risk-informed activities when supported by rigorous quantitative risk assessments and acceptable risk levels are achieved. The acceptance criteria are documented in Regulatory Guide 1.174, "An Approach for Using Probabilistic Risk Assessment in Risk-Informed Decisions on Plant-Specific Changes to the Licensing Basis" [9]. In general, the use of these criteria assumes that the risk assessments consider relevant safety margins and defense-in-depth attributes and equipment functionality, reliability, and availability. In addition, risk analyses should reflect the plant's actual design, construction, and operational practices. The risk-acceptance thresholds presented in the regulatory guide are structured as follows (see Fig. 7.2).
- Regions are established in the two planes generated by a measure of the baseline risk metric core damage frequency (i.e., the frequency of events damaging the reactor core, CDF) or large early release frequency (i.e., the frequency of events associated with radiation releases to the environment, LERF) along the x-axis. The baseline risk is the one associated with the plant operating in the "approved" condition.
- The change in those metrics (CDF or LERF) is represented along the y-axis.
- Acceptance guidelines are established for each region.

Table 7.6 Risk matrix example

		Consequences			
		Catasthrophic	Critical	Marginal	Negligible
Frequency	Frequent	High	High	Serious	Medium
	Probable	High	High	Serious	Medium
	Occasional	High	Serious	Medium	Low
	Remote	Serious	Medium	Medium	Low
	Improbable	Medium	Medium	Medium	Low
	Eliminated	Eliminated			

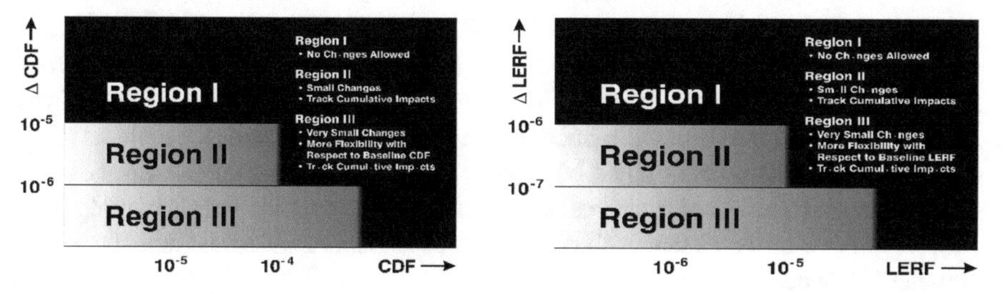

Fig. 7.2 Acceptability thresholds for CDF and LERF [9]

These guidelines are intended for comparison with a full-scope (including internal events, external events, full power, low power, and shutdown) assessment of the change in the risk metric, and when necessary, as discussed below, the baseline value of the risk metric.

Region I: Applications that result in increases to CDF above 10^{-5} per reactor year (Region I) would not normally be considered.

Region II: When the calculated increase in CDF (ΔCDF) is in the range of 10^{-6} per reactor year to 10^{-5} per reactor year, applications may only be considered if it can be reasonably shown that the total CDF is less than 10^{-4} per reactor year (Region II).

Region III: When the calculated increase in CDF (ΔCDF) is very small, which is taken as being less than 10^{-6} per reactor year, the change may be considered regardless of whether there is a calculation of the total CDF. While there is no requirement to calculate the total CDF, if there is an indication that the CDF may be considerably higher than 10^{-4} per reactor year, the focus should be on finding ways to decrease rather than increase it.

An equivalent description of these regions is also available for LERF. Notice that these thresholds are evaluated in the base plant risk and the risk change generated by a particular modification of deviation from normal operation.

7.8 F-N Curves

Another method for presenting risk is the frequency-number (F-N) curve. The F-N curve is a means of presenting societal risks, such as historical records of incidents [10]. The curve is used to plot the frequency – designated "F" for a cumulative analysis and "f" for noncumulative – of various accidents against a parameter representing a measure of the consequences, often the number of casualties associated with the accidents. Other measures of the consequences can be used to express the risk using an F-N-curve, such as the area of damage and the property loss in a monetary unit. This format can then be used to estimate the number of casualties that could be equaled or exceeded in an accident scenario and identify many small-scale accidents or a few large-scale accidents. Using an F-N curve to present the risk and use it for risk evaluation purposes is useful if the analysis consists of many scenarios. Each could be described as the risk curve representing the magnitude of the entire risk for the assessment. In Sect. 2.1, the risk was presented as a number without the possibility to distinguish between low probability/high consequences and high probability/low consequences cases. Using the same data but presenting it as an F-N curve eliminates this problem as it, in essence, is a statistical distribution of the consequences.

The F-N curve is developed as a list of all events and their associated frequencies [10]. An example of such a list is presented in Table 7.7.

A "criterion line" can be developed and superimposed on the F-N curve to provide a visual tolerance limit [10]. Often dual criterion lines are used to define an as low as reasonably practicable (ALARP) region. Mathematically, the criterion curve is expressed in Eq. 7.1.

$$F = k \times N^{-a} \tag{7.1}$$

where:

F = cumulative frequency of N or more fatalities (or monetary loss, damage area, etc.)
N = number of fatalities (or monetary loss, damage area, etc.)
a = aversion factor (commonly between 1 and 2)
k = constant

When plotted on a log-log scale, the resulting slope of Eq. 7.1 is equal to $-a$. It represents the degree of aversion to multi-fatality (or monetary loss, damage area, etc.) events represented by the criteria. Example F-N curves are presented in Fig. 7.3.

Figure 7.3a presents an example where the F-N curve passes above and below the criterion line. Passing above the criterion line represents instances where the risk exceeds the risk criterion.

Figure 7.3b presents a discretized F-N curve estimated in a manner similar to that shown in Table 7.7.

Table 7.7 F-N calculations

Event	Event frequency (per year)	Event consequence	Cumulative frequency (per year)
E_1	f_1	N_1	$F_1 = f_1$
E_2	f_2	N_2	$F_2 = f_1 + f_2$
…	…	…	…
E_n	f_n	N_n	$F_n = f_1 + f_2 + \ldots + f_n$

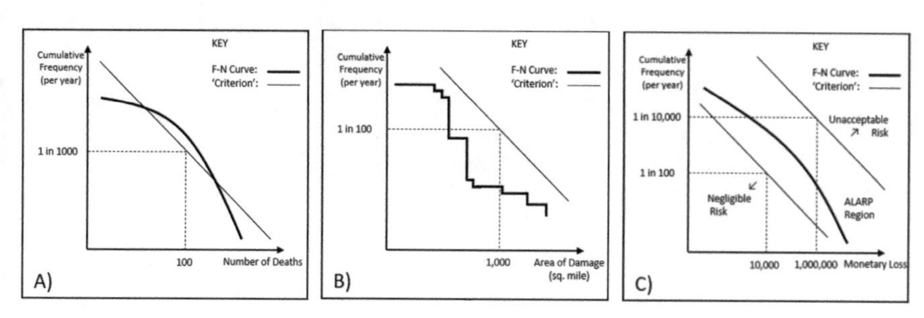

Fig. 7.3 Example F-N curves: (**a**) deaths, (**b**) damage, and (**c**) monetary loss [10]

Figure 7.3c presents an example where the F-N curve results are bound by dual criteria falling within the ALARP region.

It is worth mentioning that there are some arguments for combining the F-N curve for risk evaluation purposes as the decisions can be mathematical inconsistent in some cases; the technique is, however, frequently used in practice [11].

References

1. M. Modarres, *Risk Analysis in Engineering* (CRC Press, Boca Raton, 2006)
2. A. Wolski, J. Alston, Chapter 90, SFPE handbook of fire protection engineering, in *Fire Risk in Mass Transportation*, 5th edn., (Society of Fire Protection Engineers, Gaithersburg, 2016)
3. ISO, *23932-1 Fire Safety Engineering—General Principles* (ISO, Geneva, 2018)
4. ISO, *Guide 73:2009: Risk management—Vocabulary* (ISO, Geneva, 2009)
5. A. Wolski, N. Dempsey, B. Meacham, Accommodating perceptions of risk in performance-based building fire safety code development. Fire Saf. J. **34**, 297–309 (2000)
6. NFPA, *NFPA 551, Guide for the Evaluation of Fire Risk Assessments* (NFPA, Quincy, 2022)
7. B.F. Barry, *Risk-Informed, Performance-Based Industrial Fire Protection: An Alternative to Prescriptive Codes* (Tennessee Valley Publishing, Knoxville, 2002)
8. United States Department of Defense, *MIL-STD-882E, Standard Practice: System Safety* (United States Department of Defense, Washington, DC, 2012)
9. U.S. Nuclear Regulatory Commission, *Regulatory Guide 1.174: An Approach for Using Probabilistic Risk Assessment in Risk-Informed Decisions on Plant-Specific Changes to the Licensing Basis* (U.S. Nuclear Regulatory Commission, Washington, DC, 2011)
10. Center for Chemical Process Safety, *Appendix A, Understanding and Using F-N Curves, Guidelines for Developing Quantitative Safety Risk Criteria* (American Institute of Chemical Engineers, New York, 2009)
11. A.W. Evans, N.Q. Verlander, What is wrong with criterion FN-lines for judging the tolerability of risk. Risk Anal. **17**(2), 157–168 (1997)

Chapter 8
Fire Hazard Identification

The process of hazard identification is the first step in developing fire scenarios. The hazard identification process produces a list of distinct fire-related hazards and their specific locations that may contribute to fire scenarios. This process and the corresponding outputs (i.e., the list of hazards identified) are specific to the application (facility, process, etc.).

A "hazard" is defined as a condition or physical situation with the potential for harm. In general, hazards can relate to:

- Physical objects can be characterized by physical state (solid, liquid, or gas) or chemical properties (e.g., flammability range, volatility, density, miscibility).
- A person or a group of persons, which can be characterized in terms of human factors characterizing:
  - The response or behavior to an event (e.g., failure to evacuate within a fixed time period, limited mobility, cognitive impairments)
  - The ability to initiate an event (e.g., arson, work-related activities leading to ignition or propagation)
  - The ability to set up precursor conditions amenable to ignition and fire propagation (e.g., improper procedure adherence)
- Environmental conditions, such as lighting strikes, high winds, droughts.
- The configuration or operation of a facility (e.g., a complex building with a single exit).
- Processes or activities within a facility (e.g., a process that creates a dust explosion hazard).

There are formal recognized techniques and tools available to perform a "hazards analysis." Depending on the selected approach, the hazard analysis may include elements of qualitative risk assessment as the hazards may be classified in terms of likelihood and consequence. Detailed technical guidance on developing a hazards analysis is out of the scope of this guide. Instead, this guide highlights the importance of hazard identification as a first step in developing fire scenarios supporting

SFPE Guide to Fire Risk Assessment, The Society of Fire Protection Engineers Series, https://doi.org/10.1007/978-3-031-17700-2_8

the risk assessment. By rigorously and systematically performing the hazard analysis, the scenarios developed and evaluated should ensure:

- There is a minimal likelihood of missing a potential hazard in the assessment
- Identified hazards are not overrepresented in the assessment, which may overestimate their risk contribution

8.1 Hazard Classification

Hazards may be classified in terms of both a "hazard type" dimension and a "hazard category" dimension to assist the identification process. A "hazard type" relates to the progression of the fire scenario, while a "hazard category" refers to the human contribution or operational elements associated with the hazard.

8.1.1 Hazard Types

The term "hazard type" allows hazards to be classified with respect to the progression of the fire scenario. Specifically, the hazard type may be a (1) precursor to the event, (2) a potential ignition source, or (3) secondary (intervening) combustible hazard. Each hazard type is defined as:

- *Precursor Hazard Type:* Situations, events, or configurations that may or may not directly cause a fire but could increase the likelihood of a fire or the consequences of such an event. These are sometimes referred to as precipitating hazards. One such example is the spill of a combustible flammable liquid. Another example is the improper maintenance of a fire suppression system that fails when called upon. A wide range of hazards can fit in this category, and their connection to potential fire events may not be obvious or intuitive. Precursor hazards typically result in conditions that are unanticipated or inadequately managed by fire safety control measures put in place for normal conditions. Refer to Table 8.1 for additional examples of these precursor hazards (also called precipitating hazard types).
- *Ignition Hazard Type:* Sources can directly lead to fire event initiation. Ignition sources are those items (e.g., equipment, components) or activities (e.g., hot work activities) that can initiate a fire. Generally, any object that emits sufficient heat to ignite combustibles is a potential ignition source. These potential ignition sources can be specific to the facility or process analyzed, or the source may be generic to any facility (e.g., improper cigarette disposal). Table 8.1 includes examples of ignition hazard types.
- *Propagation Hazard Type:* Configurations that directly support the continuation or escalation of fire event consequences, that is, the rate and size the fire can grow. Fire propagation hazards are those intervening combustible materials or

Table 8.1 Hazard classification examples

		Hazard types		
		Precursors	Ignition	Propagation (intervening combustibles)
Hazard categories	Human	Improper equipment maintenance or housekeeping practices Inadequate training for handling hazardous material Cyber/internet-based activities (e.g., *sabotage*)	Incendiary devices (*eligibility of device/substance handling*) Hot work (e.g., *welding, soldering, brazing, burning*) Improper disposal of cigarettes Other activities leading to ignition	Poor storage and use practices Construction practices (e.g., *improper selection of insulating material*)
	Equipment	Tank/pipe rupture and leakage (e.g., *caused by deterioration due to corrosion*) Structural/built-in vulnerabilities (e.g., *lack of redundancy or safeguards*)	Cooking (e.g., *gas range*). Electrical failure (e.g., *short circuits, lithium-ion battery failures*) Internal combustion engines (e.g., *due to overheating*) Other hot surfaces	Combustibles (e.g., *upholstered furniture, mattresses, bedding, clothing, wood-based items*) Interior and exterior materials Leaks of combustible or flammable liquids
	Processes	Spills or unintended discharges *Lack of cleaning in HVAC ducts* (e.g., *fat buildup with cooking*) Process logic errors	Spontaneous ignition Explosions Chemicals in a reaction process	Spilled or loss of containment for flammable or combustible process fluids Combustible dust or solids (e.g., *due to failure to ventilate*)
	Environment	Seismic Flooding Drought	Lightning Static electricity Wildland fire or other exterior exposure fire	High winds Surrounding vegetation Neighboring properties/facilities Elongation distance

fuels that allow the fire and its effects to escalate. Generally, any combustible material that may be ignited when subjected to sufficient heating is a propagation hazard. Potential propagation hazards can be specific to the facility or process analyzed, or the hazard may be generic to any facility (e.g., combustibles such as paper, plastic, or insulation material). Table 8.1 includes examples of propagation hazard types.

8.1.2 Hazard Categories

The term "hazard category" refers to the human contribution or operational elements associated with the hazard. Specifically, the hazard category may be (a) human, (b) equipment, (c) processes, or (d) environment. Each hazard category can be defined as:

- Human Hazard Category: Hazards related to human behavior and interaction.
- Equipment Hazard Category: Hazards related to specific systems, equipment, or materials
- Processes Hazard Category: Hazards related to operational processes within a facility
- Environment Hazard Category: External hazards

Table 8.1 provides examples of hazards organized by type and category. This table may be used as a starting point for conducting a comprehensive hazards identification task. The different hazard types are aligned horizontally in the top row of the table. The hazard categories are aligned vertically in the first column of the table. Accordingly, the analyst may develop a table specifically for the facility or process by identifying applicable hazards within each type and category.

8.2 Tools and Techniques for Hazard Identification

There are established tools and techniques for performing hazard identification. The output of these tools and techniques is a listing of identified hazards.

These established tools can be further classified as "bottom-up" or "top-down." In a "bottom-up" approach, the hazards are identified by inspection. Once listed, potential accident scenarios are developed from the identified hazards. "What if" and HAZOP are examples of "bottom-up" approaches. In contrast, "top-down" methods start from a consequential condition, and the hazards that can produce those consequences are logically deduced. "Fault trees" are examples of logic models that are logically developed for identifying individual "failure modes" (e.g., specific hazard conditions) that can produce the consequences. The following are examples of tools that may be used:

- *Preliminary Hazard Analysis (PHA)* [1]. PHA is a "bottom-up" model, the simplest technique to identify any hazard or hazardous situation. It is based on a brainstorming activity. A PHA lists all possible events that can cause harm for a given activity, facility, or system.
- *"What If" Analysis* [2]. "What If" analysis is a "bottom-up" model, a simplified technique that involves asking what happens if a particular failure (e.g., of hardware or procedures) or event occurs. The answer will be an opinion based on the available knowledge of the stakeholders answering the question. The process can be enhanced by brainstorming among multiple stakeholders. The method seeks consistency by using standardized questions regarding practices, conditions, and failure modes of equipment. The "What If" analysis team usually includes designers and operators (including plant, process, and instrumentation) and the safety engineer/officer.
- *Hazard Identification (HAZID)* [3]. Similar to PHA, HAZID is a "bottom-up" qualitative technique for early identification of potential hazards and threats affecting people, the environment, assets, or reputation. HAZID can be used in supporting a risk assessment and also as a standalone analysis. By identifying hazards as early as possible, the likelihood and consequences of accidents are reduced and potentially eliminated through the design and operational processes. HAZID requires a balanced team of designers, maintenance engineers, system engineers, electrical engineers, quality control, operational managers, etc. The HAZID process summarizes the identified hazards ranked by likelihood and consequences of the potential accidents that they can progress to. The HAZID process can also include consideration of the mitigating strategies in place. Such strategies should be identified in the analysis to be further considered in the risk assessment.
- *Hazard and Operability Study (HAZOP)* [3]. HAZOP is a "bottom-up" tool that uses a deliberately chosen balanced team to systematically evaluate the building/facility/plant, part by part, and review how deviations from the normal design quantities and performance parameters would affect the situation. Consistency between team members is supported using standard terminology and choices for observed conditions and other variables. Appropriate remedial action is then agreed upon. A HAZOP requires a complete description of the design (up-to-date engineering drawings, line diagrams, etc.) and full working knowledge of the operating arrangements. A HAZOP is usually conducted by a team that includes designers and operators (including plant, process, maintenance, and instrumentation staff) and the safety engineer/officer.
- *Failure Modes and Effects Analysis (FMEA)* [4]. FMEA is a "bottom-up" tool where the cause of the hazard is evaluated from knowledge of equipment failure, error modes, or damage mechanisms. FMEA consists of assessing the effect of each component part failing in every possible mode. The process consists of defining the overall failure modes (usually more than one) and then listing each component failure mode that contributes to it. Failure rates are then assigned to each component level failure mode, and the totals for each of the overall modes are obtained. Additional information on the FMEA is available [5].

8.3 Fire Hazards Within a Fire Risk Assessment

From the perspective of a fire risk assessment and within a formal framework as described in the previous section, the hazard identification process should include:

- A review of relevant drawings and facility documentation as identified during the project information task of the analysis.
- A walk-through of the building or facility housing a given process or other space of concern or similar occupancies for built facilities. During this walk-through, the analyst should identify precursor, ignition source, and intervening combustible hazard types. This walk-through is often performed during the project information task, wherein the analyst observes and documents potential hazards. A walk-through may not be sufficient to identify all potential hazards because the observations made reflect a moment in time. Changes in conditions could result in additional hazards.
- For facilities in the design phase, where no walk-through is possible, an inspection of similar facilities may offer the opportunity of identifying potential hazards. This inspection should be accompanied by a review of the existing design documentation, including 3D computer models (if available) of the facility under review.
- A review of the relevant fire events in similar occupancies, facilities, or processes. These results supplement the hazards identified during the walk-through. Such a review can be integral to understanding the frequency and consequences of fires in similar facilities. However, it is essential to note that the walk-through or relevant event review may not capture all possible hazards within the scope of the assessment. The development of new materials and new technologies can create unique hazards with no previous event history.
- In conducting the hazard identification process, fire engineering judgment may be needed. Unavoidably, judgment is an element that is often necessary to supplement the review of relevant data and the walk-through of the facility under evaluation.

While the hazard identification process may vary depending on the specific application, fundamental principles should be adhered to minimize the exclusion of relevant hazards. These principles include the following:

- *Systematic approach:* The process should be thorough and consistent. It begins by identifying an appropriate starting point and continuing the inspection or review until all relevant elements are evaluated. Interactions and external factors also need to be considered. Sometimes multiple iterations might be necessary. In addition, a systematic approach may include a detailed analysis of system hardware and software, the environment in which the system will exist, and the intended usage or application.
- *Use of best available information:* It should be based on a walk-through and a review of available drawings for existing facilities. At the design stage, this might consist of the latest set of drawings. Equipment manuals and operating

instructions, data sheets and specifications, and maintenance records may also be helpful. The available information may evolve during the risk assessment and may need to be reviewed periodically or if major changes are noted.
- *Use of relevant data:* In most cases, fire events data may be helpful. This data might include fire incident data or component failure data. Data involving similar circumstances (e.g., other similar facilities or systems) can be informative, but the differences should also be considered. For example, pump failure rate data from offshore applications might provide some useful information for identifying hazards in an onshore application, but the different operating environments should be noted as a factor that may limit its applicability.
- *Comprehensive treatment:* The hazard identification process should include all applicable elements to evaluate potential scenarios. This should consist of all appropriate areas of the facility, all fire and explosion hazards, all secondary combustibles, etc. Future steps in the risk assessment process will postulate relevant scenarios incorporating the hazards identified in this task.
- *Good communication with stakeholders:* All relevant interested parties should be consulted during the hazard identification process. Stakeholders may provide different perspectives relevant to the hazard identification process that could identify additional hazards.
- *Consider future developments:* It may be necessary to foresee potential changes that will create different hazards that might not be present at the time of hazard identification. Examples are changes in fuel packages, deterioration of equipment, and occupancy changes.
- Evaluate potential hazards associated with "rare or unlikely events" (these may also be referred to as "low-frequency high-consequence events"). These hazards may lead to risk contributing scenarios that should be accounted for in the design, operation, and regulatory process.
- The hazard identification should be consistent with the scope of the fire risk assessment in terms of the physical boundaries of the facility and consideration of the period of applicability of the facility.

8.4 Example: Hazard Identification

Section A.5 presents a conceptual example of the hazard identification process.

References

1. M. Modarres, *Risk Analysis in Engineering* (CRC Press, Boca Raton, 2006)
2. American Chemical Society, Task Force of Committee on Chemical Safety, *Identifying and Evaluating Hazards in Research Laboratories, Guidelines Developed by the Hazard Identification and Evaluation* (American Chemical Society, Task Force of Committee on Chemical Safety, Washington, DC, 2015)

3. F. Crawley, *A Guide to Hazard Identification Methods*, 2nd edn. (Elsevier, Amsterdam, 2020)
4. United States Department of Defense, *MIL-STD-1629A, Procedures for Performing a Failure Modes, Effects and Criticality Analysis* (USDOD, Washington, DC, 1980)
5. ISO, *IEC 31010:2019: Risk Management—Risk Assessment Techniques* (ISO, Geneva, 2019)

Chapter 9
Fire Scenarios

A fire scenario is defined in ISO 13943 [1] as a "qualitative description of the course of a fire with respect to time, identifying key elements that characterize the studied fire and differentiate it from other possible fires." In addition, a fire scenario "typically defines the ignition and fire growth processes, the fully developed fire stage, the fire decay stage, and the environment and systems that will impact on the course of the fire."

Based on the definition of a fire scenario, the purpose of this chapter is twofold:

- Provide guidance on the process of identifying and characterizing fire scenarios included in the fire risk assessment.
- Organize the fire scenarios and their elements within the framework of the concept of risk introduced earlier in this guide (see Sect. 2.1).

The output of this task is a list of fire scenarios, perhaps in tabular form, and the corresponding qualitative characterization. This qualitative characterization may include a short narrative of the scenario including any precipitating hazards, a description of the ignition source and a general assessment of the frequency of events due to that source, the intervening combustibles to which the fire may spread, and an evaluation of the fire protection systems that may limit the consequences. This list of fire scenarios and the qualitative characterization are used in the frequency and consequence analysis sections later in this guide.

9.1 Identification and Characterization of Fire Scenarios

A fire scenario can be defined as a set of key elements characterizing a fire event. Many of these elements may have already been characterized as part of the hazard identification process described in Chap. 8. As such, the scenario identification and characterization process are intended to summarize the potential situation in a

SFPE Guide to Fire Risk Assessment, The Society of Fire Protection Engineers Series, https://doi.org/10.1007/978-3-031-17700-2_9

systematic manner that can be evaluated without overcounting or undercounting the contribution of the different hazards. The key elements in a fire scenario are:

- Ignition: Often the starting point for selecting and describing a fire scenario, that is, the first item ignited. It is also one of the elements capturing the identified hazards in the analysis. Consideration should be given to precipitating hazards and their effect on the scenario. The identification of ignition sources occurs in the hazard identification task described in Chap. 8.
- Propagation: Combustibles involved in a fire scenario other than the first item ignited. Many fire events become "significant" because of secondary combustibles, that is, the fire can propagate outside the ignition source. The identification of secondary combustibles results from the hazard identification task described in Chap. 8. As in the case of ignition, this element in the description of a fire scenario often captures the identified hazards in the analysis.
- Fire protection: Systems and barriers set in place to limit the consequences of fire scenarios. Fire protection features may include active systems (e.g., fire prevention such as control of ignition sources, automatic detection or suppression, fire dampers, smoke control strategies, egress paths) and passive systems (e.g., fire retardant materials, fire doors, firewalls, or fire-resistant structure). The effectiveness of a credited fire protection feature should be characterized in relation to:
  - Code compliance: System or fire protection feature designed, installed, and maintained in accordance with adopted/accepted code requirements for the intended hazards is understood to provide a tolerable and acceptable level of safety.
  - System impairments: Impairments are characterized in the analysis with the corresponding reliability and availability. Systems that are maintained operational and are routinely inspected and tested as programmatically required.
  - Adequacy for the identified hazard: The fire hazards analysis, supported by field walk-downs, indicates that the systems are designed and installed to mitigate the identified hazards.
- Consequences: Scenario consequences should capture the potential outcome of the fire event. It should be measured in terms of (1) relevance to the decision-making process and (2) consistency with the frequency term in the risk equation. Consequence analysis is discussed in more detail in Chaps. 10 and 11.

A fire scenario is also associated with a location which refers to both the physical location of the fire and the characteristics of the room, building, or facility in which the scenario has been postulated. In general, room characteristics include size, ventilation conditions, boundary materials, and additional information necessary for location description.

A scenario can also be quantified in terms of the likelihood of certain events occurring and the outcome (consequences) of the scenario. Still, these aspects are not part of the fire scenario per se and are dealt with in subsequent sections of this guide.

In a fire risk assessment context, an event tree is a logic model capturing the chronological events within a fire incident, representing a family (i.e., an associated group) of scenarios. In this type of model, the sequence of events is represented as a timeline as a series of branches. Each branch within the tree is characterized by its possible outcome. It is important to understand that the event tree does not represent a single scenario, but rather each individual branch (sequential pathway created by a series of branches) in the event tree is an individual scenario. Event trees represent the three key elements of the fire scenario (ignition, propagation, and fire protection features) and work forward from the initiating event (ignition) to generate branches that define a range of scenarios and outcomes resulting from secondary sub-events. Event trees are helpful when there is little data available about the outcomes of concern (e.g., low frequency/high consequence such as multiple fatality fire incidents) [2].

The extent of fire damage (i.e., the level of the consequences) associated with a fire scenario often depends on the success or failure of fire suppression actions and the characteristics of the fire if left alone. Though successful fire control or suppression increases with time, consequences usually increase with time (i.e., the longer a fire burns). Therefore, to explicitly account for detection, suppression, and containment, a number of scenarios that result in progressively larger consequences and progressively lower conditional probabilities may be modeled.

Consider, as a conceptual example, the event tree depicted in Fig. 9.1 captures the key elements of a fire scenario. In this example, scenario progression consists of ignition, fire propagation, and two suppression attempts at different times in the timeline (e.g., an automatic sprinkler system and the fire department). Depending on the success or failure of each event, four different fire scenarios occur, as represented by the consequences numbered 1 to 4 on the right-hand side of the event tree. In practice, each branch in an event tree is defined so that the top branch is a "positive" outcome scenario, and the lower branches are a combination of all "negative" or "negative" and "positive" outcome scenarios. In this example, the first fire scenario (Consequence 1) results when the fire is limited to the ignition source, that is, the sprinkler system is not activated, and the fire does not propagate to secondary

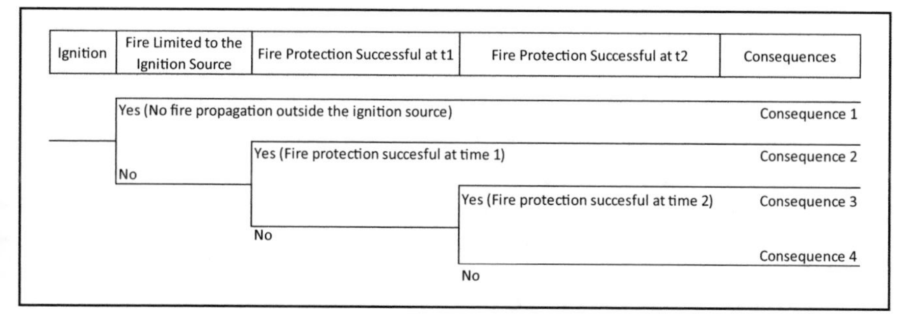

Fig. 9.1 Conceptual representation of a generic fire scenario in an event tree format

combustibles. In this instance, the consequences are likely to be minimal, especially when compared to fire scenarios 2–4 (Consequences 2 through 4).

In contrast, the fourth fire scenario (Consequence 4) occurs when the fire propagates to intervening combustibles, and both suppression attempts fail. Notice that the first event in the event tree models the likelihood of ignition. The second event models the ability of the fire to propagate, which can be represented with a conditional probability. The subsequent two events model the detection and suppression activities. These are also represented with conditional probabilities. The last event in the event tree models the consequences assigned to each fire scenario outcome.

There may be dependencies between the parameters that should be resolved to ensure correct quantification. For example, ignition should not be represented with a frequency of events that have been successfully suppressed because the subsequent conditional probabilities for suppression failure will be "double counting" the suppression credit in the analysis. To resolve this, the frequency should include all applicable fire events regardless of suppression effectiveness, as reflected in the data available for supporting the frequency estimate. Another dependency may be between subsequent suppression events. That is, the probability of successfully suppressing at Time 2 may be conditional to the suppression failure at Time 1.

It is noted that the event tree in Fig. 9.1 is generic and has been developed for explanatory purposes. Specific scenarios/applications may require an event tree with a different structure.

Since fire scenarios are the building blocks in a fire risk assessment, two key questions need to be addressed during the process of identifying fire scenarios. There is no pre-dispositioned answer to these questions as they are often answered during the process of developing the risk assessment. These two questions are:

- How many fire scenarios should be included?
- Which fire scenarios should be included?

It is noted that there are no set criteria for either of these questions. Instead, fire scenarios are selected and incorporated in a risk assessment to appropriately characterize the fire risk and meet the study's objectives. The following guidance may assist in the process of appropriately identifying the number and types of fire scenarios:

1. Use the identified hazards as the starting point for the fire scenario identification process.
   (a) Identify the initial heat sources, fuel source, and point of fire origin. This should include the initial heat source(s) continuously present in the facility or those brought in temporarily.
   (b) Assess the potential for fire growth and propagation, including the potential for secondary fuel packages.
   (c) Is there a smoldering or incipient phase? If so, what is the duration of this phase and other phases of the fire?
   (d) Does the fire reach flashover or full involvement of the first compartment or enclosed space?

(e) Does the fire spread to a second room, compartment, or space?
(f) Does the fire spread to a second floor or level?
(g) Does the fire spread beyond the building, structure, vehicle, or other objects of origin?

2. Identify the available fire safety features, including:

   (a) Fire prevention: Fire prevention features can be explicitly included in the risk assessment to reduce the ignition frequency. For example, this may involve implementing a good housekeeping program, improved testing and maintenance, or 24-h security.
   (b) Fire detection: Fire detection may include "human detection" (i.e., the ability of nearby personnel to detect the fire) or automatic systems such as smoke detection, heat detection by electronic detectors, or fusible links on sprinkler systems.
   (c) Fire suppression: Fire suppression systems may include prompt suppression activities by personnel nearby the fire (e.g., using fire extinguishers), automatic fire suppression systems, and a fire department or fire brigade.
   (d) Passive fire protection features: Passive fire protection can be included to stop or delay the progression of fire damage. Passive fire protection features may consist of interior materials, fire barrier walls or floors, fire doors, fire dampers, etc.
   (e) Means of egress: Egress routes and systems allow people to evacuate in the event of a fire. Means of egress can include the number and remoteness of exits, fire-separated egress routes, signage, emergency lighting, etc.

3. Consider elements that may impact or be impacted by human actions. Human error: Much like automatic systems, actions performed by humans are subject to failure. Therefore, consideration of human "reliability" when crediting detection or suppression by occupants should be considered.

9.2 Scenario Clusters

The process of identifying fire scenarios may generate an unmanageably large number of potentially relevant fire scenarios. Therefore, it may be necessary to group the scenarios. In general, scenarios can be grouped to form a cluster to reduce the number of scenarios in the analysis. Generally, scenarios can be combined in clusters if the consequences are similar. That is, the consequences are a common factor in the analysis that allows for the combination of individual scenarios and their corresponding frequencies. If ignition sources are grouped by consequences, their frequencies should also be combined (e.g., added). Consider as an example the analysis of a high-rise hotel. The consequences of a fire in each of the rooms on a given floor level are similar. Therefore, a scenario cluster can be developed, combining the scenarios from each room into a cluster with the frequency consisting of the sum of the individual scenarios. The fire in the room closest to the floor exit/stair (perhaps blocking it) should be considered representative of the cluster from the consequence perspective. This is the worst case for the given cluster.

Scenario clusters often provide an approximate (i.e., bounding or conservative) assessment of risk associated with the conditions captured in the cluster. This allows for an effective quantification process as groups of scenarios can be evaluated together, and risk-informed decisions can be made. Specifically:

- A scenario cluster may be found to be a low-risk contributor, and no further analysis or design changes are necessary for that group of scenarios.
- In contrast, a scenario cluster may be found to be a high-risk contributor requiring detailed evaluation of the scenarios within the cluster to identify key risk insights associated with improving fire safety.

Recall, fire scenarios are characterized by frequencies and consequences. This characterization is necessary for both qualitative and quantitative assessments.

9.3 Example: Fire Scenario Development

Section A.6 provides a conceptual example of estimating fire scenario development.

References

1. ISO, *13943:2017: Fire Safety—Vocabulary* (ISO, Geneva, 2017)
2. BSI, *PD 7974-7:2019: Application of Fire Safety Engineering Principles to the Design of Buildings. Probabilistic risk assessment* (BSI, London, 2019)

Chapter 10
Qualitative Fire Risk Estimation

Risk estimation is how the frequency and consequences for each fire scenario are developed and then combined to characterize the risk that will be used for decision-making. This chapter focuses on a qualitative approach that follows a systematic process that can be reviewed and reproduced to support risk-based decision-making for engineering solutions.

This chapter does not aim to provide a detailed description of the available frequency and consequence prediction approaches but rather an overview of the structure and examples employed in a fire risk assessment, accounting for the nature and detail of information usually collected from fire incidents. An additional listing of approaches can be found in the *SFPE Handbook of Fire Protection Engineering* [1].

The output of this process generally consists of a table or list of fire scenarios with the corresponding frequency and consequence assessments, together with their corresponding risk estimates. Depending on the objectives, the risk estimates for an individual scenario, groups of scenarios, or the entire "risk profile" for a system/facility/building can be evaluated for acceptance.

10.1 Qualitative Evaluation of Fire Risk

A qualitative assessment evaluates risk based on the merits of the specific designs versus the potential consequences of the fire events postulated. The frequency and the consequence of the fire scenario are evaluated in levels corresponding to a risk matrix. The qualitative nature of the analysis requires the identification and characterization of the factors affecting fire scenario likelihood and the corresponding consequences using a structured, systematic approach that can be reviewed, reproduced, and maintained.

SFPE Guide to Fire Risk Assessment, The Society of Fire Protection Engineers Series, https://doi.org/10.1007/978-3-031-17700-2_10

In general, for each scenario defined in the risk assessment, the analyst should:

- Qualitatively assign an ignition likelihood.
- Determine the ability of the fire protection features to limit the consequences associated with the potential outcomes of the fire scenario through the qualitative assessment of conditional probabilities.
- Assign a qualitative consequence level for each potential fire scenario.

For the family of fire scenarios represented by the event tree in Sect. 9.1, the initiating event represents ignition frequency. This initiating event should be qualitatively characterized by the frequency levels defined in the corresponding risk matrix.

The subsequent three events in the sequence are represented with conditional probabilities capturing: (1) the ability of fire to propagate outside the ignition source, (2) the ability for detection and suppression at Time 1, and (3) the ability for detection and suppression at Time 2. Since this is a qualitative evaluation, these probabilities are characterized by assessing their specific capabilities against the postulated fire scenario conditions. For example, if the ability to detect and suppress the fire is highly effective, the resulting consequences are expected to be less severe.

Finally, there are four consequence levels resulting from ignition. Each represents a potential outcome, which should be qualitatively characterized by one of the consequence levels in the risk matrix. The first consequence is associated with a fire that does not propagate. The following three levels of consequences result as the fire develops and detection and suppression attempts fail.

10.1.1 Methods for Supporting Qualitative Evaluations

The qualitative assessment of frequency, fire protection capabilities, and resulting consequences is often based on the review of fire event records, engineering judgment, or analytical modeling.

10.1.1.1 Review of Fire Events Records

Review of fire event records refers to investigating similar fire events to inform the frequency and consequence levels. This is an effective tool as it provides a realistic characterization of potential likelihood and consequences when available. Specifically, a review of the available information on loss incidents and the available loss trend data may be helpful in understanding the frequency and consequences of an incident and provides a breakdown of the resulting damage.

Fire event data may be specific to the built environment being studied (accident data from a specific operation is usually the best source of information); specific to structures of a common type sharing a common location or owner; on any larger aggregation of structures of a common type including national or international databases.

Estimating frequency and consequences using fire event records has advantages and disadvantages, including:

- *Confidence in and relevance of used data*: Fire event data can provide a basis for the values used in the assessment, that is, when sufficient relevant past data are available, historically based assessments may be adequate in making a reasonable assessment of fire risks. However, the resulting frequency and consequence assessments derived this way represent only average values and are most applicable to simple systems with few variables that can significantly change the meaningful results. The data used should be relevant to (i.e., have the same basis as) the case being studied. In addition, errors in data and changes to this data over time may be issues that can hamper the use of fire events data.
- *Accessibility of data*: The database may not be available to users, in which case it is difficult to use, and another method should be used.
- *Size of database to support precise estimates and availability of detail*: The database size is essential in estimating consequences. However, the size of the database should also be representative for accurate estimates of consequences. One of the disadvantages is the deficiency in the available data. Often, the details captured in the data do not include all the details of importance for estimating consequences.
- *Fire event databases are rarely complete:* Minor incidents, which could have escalated into major incidents, are sometimes not reported and, therefore, may not be included in the data. Consequently, the engineer should scrutinize sources of data to determine applicability.

There are many situations where fire events data may be limited or unavailable to make confident predictions about consequences. Therefore, when using fire event data, the practitioner should go through the following process:

- Compile data.
- Review and evaluate fire events data to determine the potential for fires.
- Evaluate the applicability of data. That is, determine if fire events data is relevant and appropriate for the study being undertaken.
  - If yes, then fire events data can be used.
  - If partially, then apply engineering judgment to modify the data.
  - If not, use another method.

The most important aspect of data selection is ensuring applicability and appropriateness for providing evidence and potential characteristics of the scenarios postulated in the risk assessment. Generally, in actual applications, fire frequency and consequences assessments at a specific facility or building require adjustment to fire events data to reflect the particular facility or building.

10.1.1.2 Engineering Judgment

Engineering judgment is often necessary to process the information available from empirical evidence or analytical modeling. Engineering judgment can be based on a practitioner's experience or made using a systematic and consistent procedure such as the Delphi method. This process is helpful and estimates the consequence, especially in the absence of other methods or other forms of data that are either nonexistent or lacking. It also requires skill and experience, but even experienced engineers may sometimes struggle to estimate the consequence with confidence. Another deficiency in utilizing engineering judgment is the inherent bias resulting from specific individuals, depending on past experiences.

If data is insufficient or not available, analysts will use judgment to determine baseline values, with average values being taken throughout the process, or a risk matrix can be used in which all consequence estimates are incorporated into a small number of well-distributed values. It should be noted that engineering judgment might be done for point values or ranges. Using ranges is less subject to controversy and disagreement between consequence estimating practitioners. Estimates obtained in this manner should combine the judgment and opinions of a group of engineers rather than rely on a single opinion.

Engineering judgment is an alternative or a supplement for situations in which applicable data is not fully available. Engineering judgment is based upon a practitioner's experience where other forms of data are either nonexistent or lacking. Engineering judgment, however, has drawbacks:

- Individuals may have an inherent bias dependent upon experience. Utilizing expert elicitation procedures such as Delphi panels can minimize individual biases by using a panel of opinions rather than relying on a single opinion.
- Individuals may underestimate low frequencies while overestimating high frequencies.
- Individuals may misestimate unique or high hazard events, treating them as impossible (negligible frequency) if they have never occurred and as more likely than they are if they have occurred, particularly if they have occurred recently.
- Individuals may treat conditions that are not independent as independent (i.e., treating conditional probabilities as unconditional).
- Redundant systems may not significantly increase reliability as much as individuals assume. For instance, even where multiple sprinkler systems are installed, an inadequate water supply or improper maintenance techniques may compromise the operation of all systems. System reliability can be a complex function of component reliability, and individuals may not be equally skilled at estimating human error and mechanical reliability.

The following techniques can be used to improve judgments:

- *Ranging.* A common approach is to have the panel establish a best estimate value or position and then assess upper and lower bounds. While this is the most common approach, ranges established using this technique often underpredict reality (see the following technique for a better approach).

- *Bracketing*. Often it is challenging to select the best and most representative value. Some analysts have found that estimating the extreme values (high and low) is more manageable. Once these extreme values are available, selecting the most representative value is usually easier and often more defensible.
- *Partitioning*. When a value is difficult to establish directly, segmenting the problem into parts can make the situation more tractable. This approach is commonly used in selecting event frequencies (e.g., event trees). Careful selection of the segments can usually make the analysis more defensible.
- *Iteration*. For some problems, the numerical answer is less important than the conclusion relative to the tolerable risk. In such instances, approximate solutions can reinforce judgment.

10.1.1.3 Analytical Modeling

Analytical modeling is used for the most part to assess the consequences of a specific fire scenario beginning in a particular location. The results are the number of deaths and injuries, cost of property damage, interruption to business operations or downtime, or the environmental impact.

Deterministic models have been developed and are continuously being refined and validated to estimate consequences or carry out consequence analysis. The available consequence computer models generally include the capability to evaluate fire development, smoke movement, structural response, and response and evacuation times. They also estimate time to critical damage thresholds and untenable conditions.

An advantage of using models to evaluate consequences is that they provide a quantitative estimate based on a rationalized method. In addition, any change in the design can be logically related to the resulting consequence. This allows designers to easily identify where to make changes to produce acceptable fire risk estimates.

When using models, the users should be aware of their limitations so that their application does not compromise the resulting consequences. The inputs to the models can also be a concern as the data fields may be subjective, based on judgment, or difficult to obtain. At the same time, uncertainty should be considered (see Chap. 13).

The models used in evaluating fire consequences may be simple correlations, separate individual models, or a complete analysis combining all the required models. The usage of one method or another depends, in part, on the complexity of the problem being studied and the outcome sought from the study. Simple correlations are easy to use but may not provide the required output and may be combined with other models. Individual separate models can give the resulting outcome. Still, the user may have to feed the output of one model to another and ensure that the limitations of all models are well understood. In a complete analysis combining models, the user does not need to worry about the links between the different models;

however, the input data, which is a long process, should be well prepared and accurate. The *SFPE Handbook of Fire Protection Engineering* provides additional information on computer simulation and risk assessment [2].

10.1.2 Qualitative Analysis

As noted in Sect. 3.7, the frequency analysis characterizes how often the scenarios, including their potential consequences, may occur (per unit of time for frequency). Previous steps in the fire risk assessment have identified ignition hazards and defined scenarios capturing such hazards. Each ignition source included in the fire scenarios should be assigned a preliminary frequency level according to the risk matrix defined for the assessment.

Recall that Chap. 8 summarized typical ignition hazards. These ignition hazards are broadly classified as human, equipment, processes, and environmental.

- Ignition hazards in the "human" category are associated with sources that are not typically present in the area where the fire occurs and are brought into the area temporarily. Factors affecting the likelihood of these ignition sources include the level of occupancy, maintenance or hot work activities, and storage level in the area. It is recommended that each of these factors be evaluated when assessing the frequency.
  - A fire hazards analysis should provide a listing of ignition sources and combustibles stored in the area.
  - Maintenance and hot work records, together with fire events records in the facility, often assist in the determination of ignition frequency levels.
  - The level of occupancy (i.e., type and magnitude of the occupancy) can influence ignition as high occupancies may increase the likelihood of ignition.
- Ignition hazards due to "equipment" refers primarily to components, machinery, or other permanently located items in the area where the fire is postulated. Fire event records associated with fire events in the equipment (or similar equipment) often provide the necessary information for a qualitative assessment of ignition frequency.
- As discussed above, due to the "processes" classification, ignition hazards can be comprehensively accounted for by subdividing them into equipment and human actions necessary for process functionality.
- Environmentally induced ignition hazards are those not generated by humans, equipment, or processes. A review of weather patterns and fire event records in the geographical area may provide information necessary to inform the qualitative frequency assessment.

The above classification assists in identifying ignition sources. Once identified, ignition sources are, in practice, associated with either fixed ignition sources or "transient" ignition sources to assign frequencies. Fixed ignition sources are those permanently located in the facility within the scope study. On the other hand, transient sources represent ignition hazards or combustibles found in the facility temporarily. Examples of transient sources may include ignition sources brought into the facility by visitors, plant personnel during maintenance activities.

Fixed ignition sources are assigned frequencies representing the likelihood of the respective source. In the fire risk assessment, that frequency is the source of fire risk contribution at the specific location of the ignition hazard (i.e., where the fire scenario has been identified). In contrast, transient sources may not have a particular area. Instead, the analyst is expected to include transient fire scenarios in appropriate locations within the scope of the assessment. The selected areas are often based on the postulated transient scenario's ability to propagate and generate consequences. Under this approach, the analyst evaluates all potential locations where transient fires could occur and "screens out" of the analysis those with negligible consequences.

Consider a scenario associated with hot work in a facility protected by automatic sprinklers as a conceptual example. Procedures are in place requiring a fire watch to be posted while hot work activities are ongoing. Figure 10.1 conceptually captures the event tree as follows:

- The first event is fire ignition.
- The second event, "Fire Limited to Ignition Source," refers to the fire propagating outside the ignition source. In this scenario, the event consists of welding sparks or slags igniting nearby combustibles.
- The third event, "Fire Watch Intervention," refers to prompt actions by the fire watch to control the fire and prevent further propagation. This suggests that the fire at the time of this event is relatively small to be handled by a fire watch using a fire extinguisher.
- The fourth event refers to a fire that the fire watch failed to control, continued growing, and was detected and suppressed by the automatic sprinkler system before flashover conditions occurred.

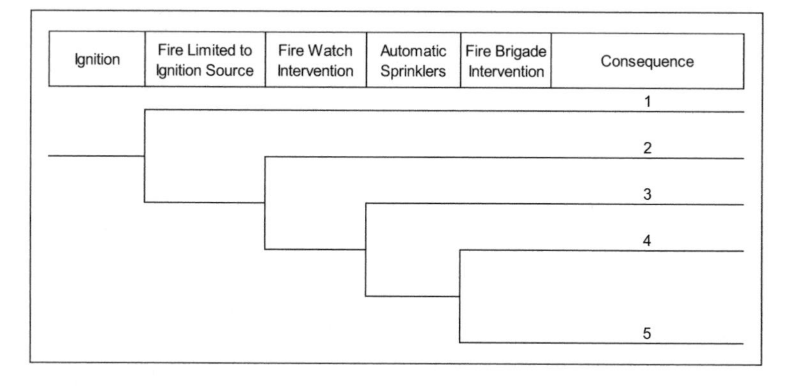

Fig. 10.1 Scenario progression event tree for hot work example

- The fifth event, "Fire Brigade Intervention," refers to a fire that the automatic sprinkler system failed to control, and the fire brigade had to intervene before flashover conditions occurred. If the fire brigade is not successfully controlling the fire, a flashover is expected to occur.

The event tree results in five potential fire scenarios and consequences labeled one through five in Fig. 10.1. The first scenario is associated with a fire that does not propagate outside the ignition source (e.g., a welding slag generates no ignition). In contrast, the last fire scenario and consequence level are associated with an uncontrolled fire, possibly flashover conditions.

Based on a review of the fire incident reports in the hypothetical facility and the relatively large number of hot work operations performed, this ignition source is ranked as "probable." That is, fire events due to hot work have been experienced or recorded several times throughout the facility's lifetime. It is noted that the characterization of "probable" is limited to the ignition event, regardless of the consequences these events have generated in the past. In that way, the frequency characterization is independent of fire protection features or mitigative strategies explicitly credited later to reduce the fire scenario frequency.

Figure 10.2 below depicts the qualitative frequency assessment of ignition (by hot work) for this example. The qualitative assessment of the frequency is "probable."

This conceptual example highlights the use of data and engineering judgment for supporting the qualitative ignition frequency assessment. Specifically:

1. Although a numerical value is not calculated, the qualitative assessment should capture the likelihood of ignition as a range of potential values. In this case, the assessment is influenced by the number of hot work activities in the facility and fire events reported in the past resulting from those activities.
2. The assessment should not be limited to facility data as such events "may not have happened in the facility under evaluation." Research outside of the facility or process under study is recommended for identifying similar events in similar industries.

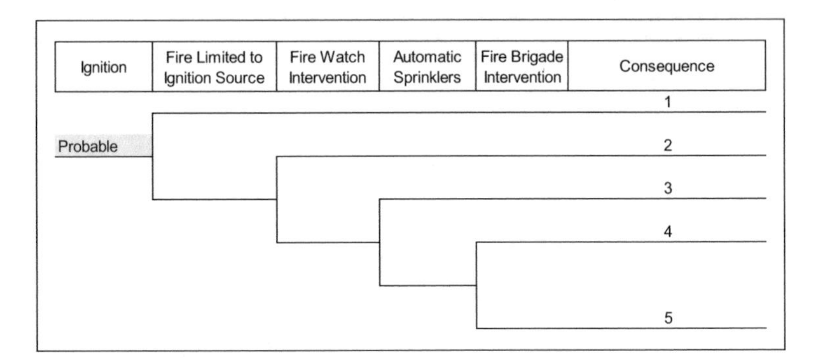

Fig. 10.2 Scenario family progression event tree for hot work example with assessment for ignition frequency

3. The frequency level determination should remain independent of the ability to control the fire (e.g., detect and suppress the fire) and the resulting consequences, as these are assessed individually as part of the risk assessment.

10.1.3 Qualitative Assessment of Conditional Probabilities

Recall that conditional probabilities are factors representing specific elements of a fire scenario, such as fire protection features. This specifically refers to the effectiveness, reliability, and availability of procedures or systems to perform as designed in the context of the postulated fire scenarios. The concepts of effectiveness and availability are typically and explicitly covered by code requirements. Applicable codes include specific requirements to ensure that the system mitigates the hazards for which it was designed and requirements for inspection, testing, and maintenance of such systems. Therefore, a code-compliant facility can be associated with tolerable risk levels contingent on maintaining the occupancy and hazards for which the fire protection program was designed to mitigate. In a qualitative evaluation, the concept of code compliance covering effectiveness and availability provides the technical basis for characterizing fire protection features in the risk assessment.

Effectiveness refers to the ability of the system or procedure to mitigate the fire hazard. Typically, a "code compliant" system would be expected to mitigate the fire hazards for which it was designed if occupancy and utilization of the facility have not changed or the fire risk assessment has identified hazards for which the system was not designed. The concept of effectiveness also refers to the system's ability to respond promptly to prevent or control the consequences of the fire scenario.

For systems, reliability and availability refer to the proper maintenance, testing, and inspections to increase the probability of success in on-demand operation (i.e., reduce the failure probability and the unavailability of the system on demand). Related to the concept of system availability, the analyst may consider compensatory measures set in place for when fire protection systems are out of service. For procedures followed by personnel, for example, facility staff, fire brigade, fire department, the concept of reliability and availability are established based on routine training.

Several methods are available to systematically qualitatively model conditional probabilities capturing the systems' effectiveness, reliability, and availability. One of such methods is the Fire Safety Concepts Tree (FSCT, NFPA 550, see Fig. 10.3) [3]. The FSCT provides a comprehensive structure for analyzing the potential impact of fire safety strategies for the different fire scenarios included in the risk assessment. The term "comprehensive" refers to a method that explicitly considers all relevant elements in a fire protection program or strategy. From that perspective, it minimizes the potential of an incomplete assessment.

In the FSCT, a logic structure is developed under the top element. The logic structure is a "success tree," which refers to logic capturing a strategy to meet the fire safety objectives (see Fig. 10.3). The fire safety objective may be as broad as

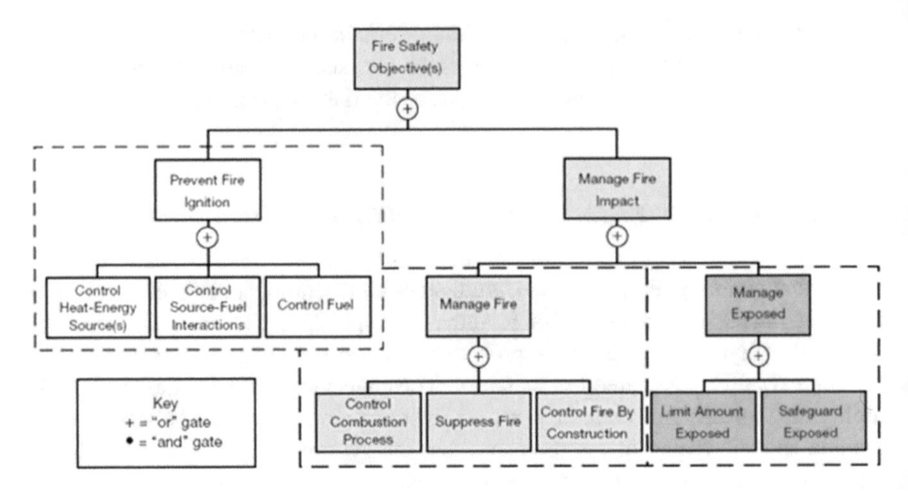

Fig. 10.3 Fire Safety Concepts Tree: top gates (NFPA 550) [3]

protecting business continuity for an entire facility or as focused as protecting a single target piece of equipment from a localized ignition source. In general, the idea is to link the elements defining a fire scenario (described earlier in Chap. 9, Fire Scenarios) to the elements in the FSCT. As a "success tree" logic model, the FSCT is static in time. It does not capture the dynamic nature of a fire scenario, which, as discussed earlier, but includes a chronology of events in time. Consequently, it may be necessary to evaluate different points in time to determine the risk of each scenario.

The two top branches in the FSCT are *Prevent Fire Ignition* and *Manage Fire Impact*. The *Prevent Fire Ignition* branch is directly associated with fire ignition frequencies by addressing the control of combustibles and ignition sources. The *Manage Fire Impact* branch includes factors related to managing the fire itself in terms of controlling the combustible material available, detection and suppression features, and managing the exposure to fire.

The logic in the FSCT is developed based on the "success" of the main safety objective. This introduces the concept of a "Path Set" for evaluating the tree. A "Path Set" is defined as the combination of events in the tree through the logic gates (enabling access to the top, indicating that the fire safety objectives are met).

For example, only one path set available that reaches the top of the tree is an indication that fire safety objectives are achieved (See Fig. 10.4a) with no redundancy or diversity in the overall strategy. On the other hand, the availability of multiple paths suggests a balanced strategy allowing more than one alternative to meet the fire safety goals (See Fig. 10.4b). By progressively moving through the various elements in the tree logically, all aspects of fire safety are evaluated to demonstrate how each may influence the objectives.

The lowest level elements are inputs to an "OR" gate. If at least one branch of the lowest level elements is considered "true" based on the designed fire safety system,

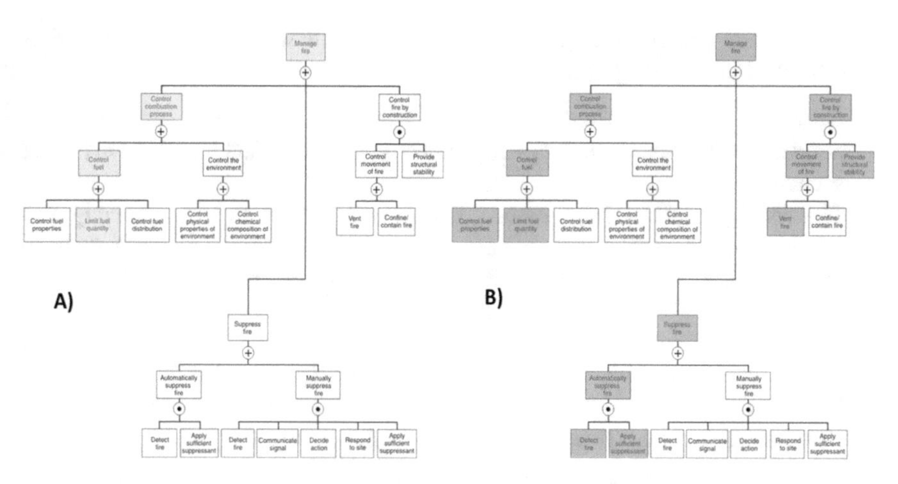

Fig. 10.4 (**a**) Minimal achieved path set, (**b**) multiple achieved path sets

the next level is considered adequate. As the evaluation is performed, each element in the lowest level of the tree is assessed from the perspective of frequency and consequences based on the categories identified in the applicable risk matrix. NFPA 550 [3] recommends the following classifications for evaluating the FSCT, which can support the qualitative risk assessment of fire scenarios.

- *Nonexistent:* A nonexistent classification is applied if there are no procedures, systems, or strategies to support the fire safety element in the tree.
- *Below Standard:* This classification refers to the fire protection systems or strategies affected by noncompliance to the applicable code or standard.
- *Standard:* This classification refers to the fire protection systems or strategies that meet the applicable code or standard for the associated occupancy and hazards.
- *Above Standard:* This classification refers to the fire protection systems or strategies that exceed the applicable code or standard for the associated occupancy and hazards.
- *Not Assessed:* This classification is not explicitly included in NFPA 550 [3] but is included in this study to identify the fire safety concept tree elements that are not assessed in the frequency or consequence assessment.

By systematically evaluating each element of the FSCT, which represents the capabilities of the fire protection program for minimizing potential consequences postulated in the fire scenarios included in the fire risk assessment, a qualitative characterization can be included in the evaluation.

Consider as an example the case of fire prevention strategies incorporated in a fire protection program. The FSCT provides a qualitative means of systematically estimating critical elements associated with ignition and fire propagation through the *Prevent Fire Ignition* logic structure of the success tree. This portion of the tree considers three primary branches, which directly relate to fire prevention practices influencing ignition and propagation:

- *Control Heat-Energy Source(s)*: This branch assists in characterizing the likelihood of controlling the heat source that would ignite the fuel identified in the fire scenario.
- *Control Source-Fuel Interactions*: This branch assists in characterizing fuel interactions – examples include barriers or spatial separations between the heat-energy source and the fuel and limiting the means of propagating that energy from the source to the fuel (by conductive, convective, or radiative means).
- *Control Fuel*: This branch assists in characterizing methods for controlling the fuel available for combustion in the fire scenario. Specifically, the ignitability – by controlling fuel properties or the environment – and the likelihood of entirely eliminating the fuel are captured in this branch.

Continuing with the example, it is assumed that inspection and review of procedures (e.g., fire prevention procedures, hot work fire watch), the analyst can rank this factor consistent with the categories in the selected risk matrix. NFPA 550 [3] provides a simple representation of this process in the *Prevent Fire Ignition* path for a computer facility recreated below in Fig. 10.5.

In Fig. 10.5, each of the events is estimated using the *Standard (S)* and *Nonexistent (N)* rankings, with the likelihood estimation that the prevention of ignition or propagation is equal to that associated with systems and strategies that meet all applicable codes and standards for the postulated fires in the risk assessment (i.e., in this case, no fire hazard was identified and included as part of the selected scenarios that would challenge the ability of the fire prevention measures to perform as designed). This estimation is made even though the control on interactions between the heat-energy source and fuel are nonexistent. The OR gate under Prevent Fire Ignition allows for prevention to be met or successful if any single branch is successful. In

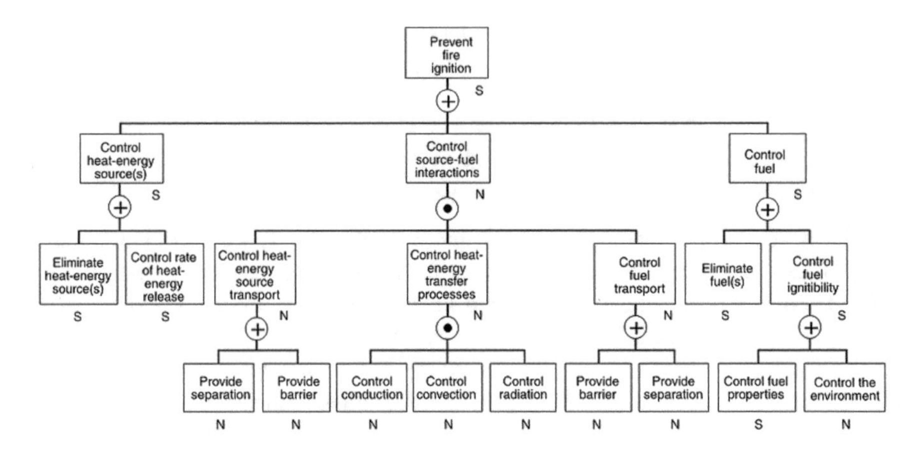

Fig. 10.5 Fire prevention in a computer facility (NFPA 550) [3]

this specific example, two paths are classified as "Standard" leading to the "Prevent Fire Ignition Branch." Once the fire protection capabilities are characterized in the context of each scenario included in the fire risk assessment, they can be qualitatively incorporated into the analysis to represent conditional probabilities.

In practice, the qualitative nature of the analysis requires that the scenario sequence be estimated systematically by determining the effects of each fire protection feature included has in the ignition frequency and resulting consequences. Each event in the tree is associated with a split fraction that eventually leads to a consequence. Since the split fractions represent fire protection features, the analyst assesses how the scenario frequency is maintained at the level of the previous event in the chronology or reduced to reflect the successful operation of the fire protection feature. In the hot work example (see Fig. 10.2), consider the case in which all the events in the chronology are deemed ineffective, resulting in an ignition that will propagate and will not be suppressed by the fire watch, the automatic sprinklers, or the fire brigade.

In contrast, a scenario in which all the fire protection features are fully effective will result in "Negligible" consequences, which result in "Acceptable" risk estimation. These two situations represent the extreme cases in which fire protection features are treated as fully effective or ineffective, which is not consistent with the probabilistic treatment of these features in a risk assessment. Once each fire protection feature is assessed with a failure likelihood assessed qualitatively, each scenario can be characterized to reflect such likelihoods. In a fully qualitative assessment, there is no specific method to capture this likelihood. However, the following concepts can provide a frame of reference for determining fire scenario frequency levels:

- *The concept of code equivalency.* Can be used to support the assessment. For example, full code compliance often suggests acceptable or tolerable levels of risk, once it is established that the systems are designed for the hazards identified and captured in the postulated fire scenarios. This is because, in most situations, AHJs accept the level of fire safety associated with code compliance facilities.
- *The concept of availability.* In this context, availability refers to the fraction of time the fire protection feature is available for "on-demand" operation. Such fraction can suggest the impact the given element may have in further lowering the fire scenario frequency and resulting consequence.
- *The concept of time.* The capability of the fire protection features to provide timely protection can also suggest the impact it may have on the fire scenario frequency and resulting consequence. A feature deemed not to have a timely response should not be credited for risk reduction.

The qualitative evaluation of conditional probabilities is perhaps best described as part of the key element depicted in an event tree.

Conditional probabilities for fire severity: The second event in the event tree presented in Fig. 10.2, "fire limited to the ignition source," is the first conditional probability in the analysis. It represents the fire's potential to propagate outside the ignition source. In the FSCT, the severity concept may be captured under the *Manage Fire* branch (see Fig. 10.4). The path under this branch includes *Control of the Combustion Process* and *Control of Fire by Construction*. The factors associated with these branches include:

- Control fuel properties
- Limit fuel quantity
- Control fuel distribution
- Control physical properties of the environment
- Control chemical composition of the environment
- Confined/contain fire
- Vent fire
- Provide structural stability

Each of these factors influences the likelihood a fire could grow into a severe event. In summary, the rankings of the fire severity are driven by the development of the fire scenarios as captured in the event tree. As the fire grows and develops, its ability to propagate and overcome existing barriers (i.e., fire protection features) can lead to different consequence levels that eventually, together with the frequency assessment, will characterize the overall risk of the various scenarios.

Using the example presented earlier on the FSCT to a fire scenario initiated by hot work activities, a "standard" classification is assigned to the fire prevention capabilities (for this example). This is based on the fire protection program at the facility, which requires fire prevention measures in place during hot work operations, including the presence of a trained fire watch equipped with fire extinguishers and the use of fire blankets (i.e., welding blankets) for protection of nearby combustibles. Notice that this is directly related to limiting the number of exposed combustibles as suggested in the FSCT under the "Managed Exposed" event. Based on this evaluation, the top branch in the event tree in Fig. 10.6 will have a relatively high probability of representing a "standard" evaluation resulting from the FSCT. The practical implication of this qualitative assessment is that the success branch, qualitatively characterized as standard, has been evaluated for effectiveness and availability and judged to have a high probability of success.

Consequently, the top branch remains with a "probable" classification as a relatively high probability has the effect of not lowering the classification. The failure branch of this event leads to Scenario 2 through Scenario 5 (Consequence 2 to Consequence 5) and should have a low probability associated with the failure of a fire prevention capability that has been evaluated as "standard." This suggests that the "probable" classification can be lowered to "occasional." The classification of "occasional" serves as the "entry point" for the next event in the chronology, which is "fire watch intervention." Notice that most of the probability will be apportioned to Scenario 1 (Consequence 1″), which represents fires limited to the ignition source with no propagation. This example illustrates how the systematic evaluation of fire protection capabilities using the FSCT directly connects with the fire scenario and informs the resulting fire scenario consequence level. Figure 10.6 illustrates this assessment.

Conditional probabilities for detection and suppression: Continuing with the analysis of the fire scenario progression event tree described in Fig. 10.6, the events

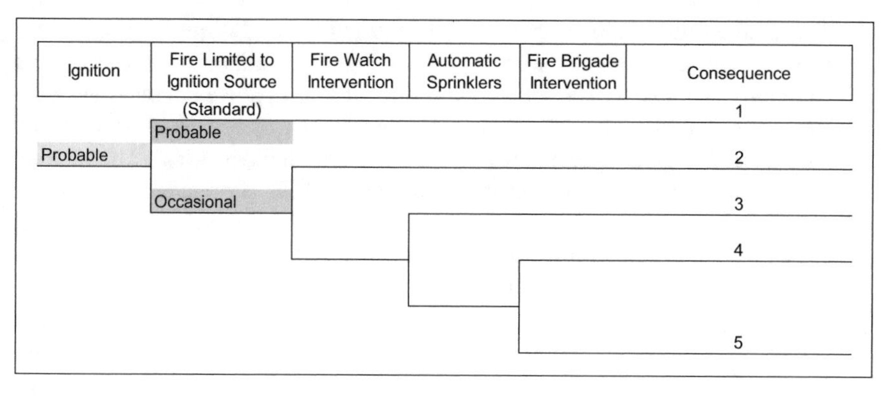

Fig. 10.6 Event tree for hot work example with assessment for fire prevention measures

occurring after "Fire Limited to the Ignition Source" are "Fire Watch Intervention," "Automatic Sprinklers," and "Fire Brigade Intervention." These three events can be interpreted as capturing the ability to detect and suppress the fire at different chronological points in time. The analyst selects the points in time to represent a stage in the fire progression leading to a different consequence level. For example, automatic sprinkler protection may be selected as the time to flashover for a room. Accordingly, failure to detect and suppress the fire at the time of flashover will likely result in a higher level of consequence if this room is equipped with sprinklers; it is likely that detection and suppression will have occurred before the fire generates conditions conducive to flashover. On the other hand, if the facility relies on a fire brigade or the response of a local fire department instead of automatic sprinklers, their corresponding average response times will need to be compared to the estimated flashover time before determining the likelihood of a timely response. It is noted that such a systematic evaluation is recommended for each time step modeled in the scenario progression.

Fire detection and suppression elements in the FSCT are found under *Suppress Fire*. These include both manual and automatic fire detection and suppression capabilities. The characterization process includes the evaluation of:

1. Automatic suppression:
   (a) Automatic detection
   (b) Application of sufficient suppressant agent
2. Manual suppression:
   (a) Detection/notification capabilities other than by manual or automatic systems
   (b) Time to respond to the site following fire detection
   (c) Decide action
   (d) Application of sufficient suppressant agent
   (e) Communication capabilities

As discussed earlier in this chapter, the fire protection capabilities included in the fire scenarios are evaluated for effectiveness, reliability/availability, and timeliness of operation. The analyst should rank these factors as applicable consistent with existing code compliance evaluations, field inspection, or review of design and operational drawings or procedures (e.g., fire prevention procedures). The ranking may use the assessment levels recommended for the FSCT. The element of response time should also be incorporated in the analysis. For a fire protection feature, it not only needs to be designed for effectiveness and maintained to be available upon demand, but it should also be capable of performing its function before the postulated consequences are realized.

Continuing with the hot works example, the fire watch and the automatic sprinklers are ranked as "standard" in Fig. 10.7. In practice, they are designed for the hazard, available to operate upon demand, and expected to respond in time.

It is reasonable to expect the fire watch and sprinkler systems to activate before flashover conditions as activation temperatures for the sprinklers are significantly lower than those observed in flashover conditions. Therefore, suppression can be included in the analysis as lowering the risk associated with flashover consequences. On the other hand, even if a fire brigade or fire department capability is evaluated as "standard," the analyst will need to determine if it will respond in time for the specific hazard modeled in the scenario. This example assumes that the fire brigade response time may not be fast enough to prevent flashover conditions. This is ranked as "Delayed" in Fig. 10.7. Notice in the event tree depicted in Fig. 10.7, that the scenario frequency classification is maintained or lowered as appropriate in each branch to incorporate the qualitative assessment of the fire protection capability in the analysis. The fire watch and the automatic sprinkler capabilities are lowered to reflect the expectation of effectiveness, availability, and timely response. Since the fire brigade is not expected to provide a timely response to prevent flashover, its entry condition of "improbable" is maintained.

In summary, each detection and suppression capability included in the analysis should be evaluated for effectiveness (i.e., it is designed and installed for the postulated hazard), availability (it is inspected and tested routinely to ensure a high

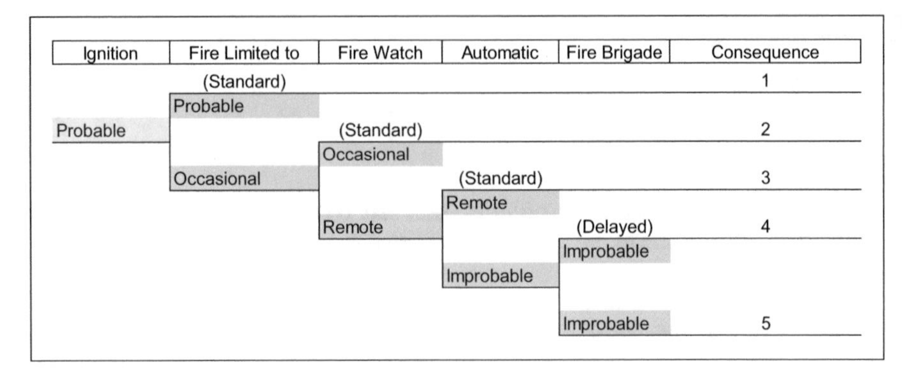

Fig. 10.7 Event tree for hot work example with assessment detection and suppression capabilities

probability of operation upon demand and is maintained in service), and timely response to protect the postulated consequences. The evaluation results for these three elements should suggest a qualitative assessment of the likelihood of successful operation in time.

Conditional probabilities for failure to operate manual systems: This probability includes two influencing factors: the reliability and availability of the hardware (i.e., the system itself) and the ability of personnel to activate the system on time. The former refers to the sum of the system's unreliability and unavailability discussed earlier in this chapter. The latter can be calculated using human reliability techniques. Human reliability analysis refers to the field of reliability engineering focused on quantifying probabilities of human errors. This is often used in engineering analysis for quantifying the probability of operator errors performing actions necessary for the successful operation of a system. This approach is scenario-specific and provides a structure for analyzing the potential failure modes associated with completing the action in detail. In the context of this application, human reliability analysis can be used to assess the ability to activate a manual system in time.

In general, the calculation of human error probabilities using human reliability analysis techniques has a qualitative and a quantitative step. The qualitative step, also referred to as feasibility analysis, consists of evaluating a set of performance shaping factors for determining if the action to be performed by personnel is likely to fail (or if the action can be performed with high reliability). Performance shaping factors are interdependent, and their impact on resulting human error probability can be challenging to assess. However, for practical analysis, the elements are often treated independently. If the qualitative evaluation suggests that the action is feasible (i.e., it can be performed with high reliability), the quantitative step is then implemented for quantifying a human error probability. Alternatively, if the qualitative evaluation suggests that the action is not feasible, the quantitative assessment should determine a human error probability close to 1.0.

The qualitative step is recommended for determining if the action is feasible and consists of evaluating a set of performance shaping factors. Focusing independently on each factor provides a structured approach for identifying essential elements in the activities that need to be completed, which identifies procedures, improvements, and training. The following performance shaping factors are often considered [4].

- ***Available Time:*** This refers to the amount of time available to diagnose and act upon an abnormal event. A shortage of time can affect the ability to think clearly and consider alternatives. It may also affect the ability to perform or complete the activity. In the context of activating a manual fire protection system, the action may be analyzed as follows:
  - The time available for performing the action may be the time between fire detection and the time for the system to be activated before the consequences occur (i.e., the time defined in the event tree for each scenario).
  - The cognition time refers to the time it takes personnel to react to an alarm and decide on the course of action.

  – The execution time refers to the time it takes to act once it has been decided on the course of action.

- The cognition and execution times, when added together, should be shorter than the time available for performing the action. Generally, if these are similar, there will be no margin in time to perform the action comfortably, increasing the resulting human error probability.
- ***Stress/Stressors:*** Stress has been broadly defined and used to describe the negative and positive motivating forces of human performance. Stress can include mental stress, excessive workload, or physical stress (such as imposed by complex environmental factors). Environmental factors often referred to as stressors, such as extreme heat, noise, poor ventilation, or radiation, can induce stress in a person and affect the operator's mental or physical performance. In the context of the failure of combustible controls as a pre-initiator event, the analyst should consider implementing the combustible control procedures as this is often part of an ongoing fire prevention program. Therefore, stress, which is a critical factor in post-initiator human response actions to accidents, may not affect the feasibility of implementing combustible control procedures.
- ***Complexity:*** Refers to how difficult the task is to perform in the given context. Complexity considers both the task and the environment in which it is to be performed. The more difficult the task is to perform, the greater the chance for human error. Similarly, the more ambiguous the task is, the greater the chance for human error. Complexity also considers the mental effort required, such as performing mental calculations, memory requirements, understanding the underlying model of how the system works, and relying on knowledge instead of training or practice. Complexity can also refer to physical efforts required, such as physical actions that are difficult because of complicated patterns of movements. Complexity can be a factor in the context of the failure to implement combustible controls. There may be scenarios in which the configuration or the specific conditions of a postulated scenario overcome typical fire prevention practices available in the facility. Fire scenarios associated with arson, for example, may bypass such combustible controls.
- ***Experience/Training:*** This performance shaping factor refers to the experience and training of personnel assigned to perform the task. Included in this consideration are years of experience of the individual or crew and whether or not they have been trained for the postulated scenario in the analysis. Another consideration is whether or not the scenario is novel or unique (i.e., whether or not personnel has been involved in similar scenarios, in either a training or operational setting). In the context of conditional probabilities for modeling fire watch response in a fire risk assessment, this factor refers to the formal training given to the fire watch before performing his duties.
- ***Procedures:*** Refers to the existence and use of formal procedures for the tasks under consideration. Common problems in event investigation procedures include situations where procedures give wrong or inadequate information regarding a particular control sequence. Another common problem is the ambi-

guity of steps. In the context of conditional probabilities for modeling fire watch response in a fire risk assessment, procedures refer to the set of instructions given to the fire watch to perform his task.

- ***Ergonomics and Human Machine Interface:*** Ergonomics refers to the equipment, displays and controls, layout, quality and quantity of information available from instrumentation, and the interaction of personnel with the equipment to carry out tasks. Aspects of human-machine interactions are included in this category. The adequacy or inadequacy of computer software is also included in these performance shaping factors. This performance shaping factor is not generally applicable to fire watch activities.
- ***Fitness for Duty:*** Fitness for duty refers to whether or not the individual performing the task is physically and mentally fit to complete the task at the time. Factors that may affect fitness include fatigue, sickness, drug use (legal or illegal), overconfidence, personal problems, and distractions. Fitness for duty includes factors associated with individuals but not related to training, experience, or stress.
- ***Work Processes:*** Refers to doing work, including interorganizational, safety culture, work planning, communication, and management support and policies. How work is planned, communicated, and executed can affect individual and crew performance. If planning and communication are lacking, then individuals may not fully understand the work requirements. Work processes include consideration of coordination, command, and control. Work processes also include any management, organizational, or supervisory factors that may affect performance.
- ***Available Staffing Resources:*** A fire accident can introduce additional demands for staffing resources beyond what is typically assumed for handling accidents not involving fire. These demands can take the form of needing to use and coordinate with more personnel such as the fire brigade and local fire department personnel.
- ***Environment in Which the Act Needs to Be Performed:*** Fires can introduce new environmental considerations not typically experienced in response to internal events. These include heat, smoke, the use of water or other fire-suppression agents or chemicals, toxic gases, and different radiation exposure or contamination levels. Any or all of these may affect the accomplishment of the desired action.
- ***Accessibility and Operability of Equipment to Be Manipulated:*** Fires and their effects (e.g., environment) could eliminate or delay the ability to take actions otherwise credited in other types of accidents because the location is inaccessible. Additionally, fires can cause the failure of equipment used in the desirable action (e.g., irreversible damage), so it should be considered inoperable, even manually.
- **T*he Need for Special Tools (Keys, Ladders, Hoses,* Etc.*):*** Fires may cause the need for special tools or clothing (e.g., breathing gear, protective clothing). The accessibility of these tools or clothing needs to be checked so that the desired actions can indeed be performed in a fire situation. Furthermore, the level of familiarity and training in using these special tools need to be assessed.

- ***Communications***: Necessary communications to carry out the desired actions may or may not be available in some fires. This needs to be checked, as does the level of familiarity and training to use any special communication devices.

The feasibility of an action is determined by evaluating each of the factors listed above against the activities required to perform the action. It is noted that some of the factors may not be applicable. The main objective in the evaluation is to identify if any of these actions would prevent the action from completion in time, which suggests a failure probability of 1.0, which qualitatively can be expressed as the action is not expected to be successfully completed.

Modeling egress in a fire risk assessment: The conceptual example associated with hot work hazards discussed so far has focused on detection and suppression capabilities represented chronologically in the event tree. Egress capabilities can be implicitly evaluated as part of the chronology of the event tree by expressing the occupant's ability to evacuate the facility in the consequence term. Each point in time captured in the chronology of the tree is associated with a fire size that will challenge the occupant's ability to evacuate. For each postulated fire scenario, egress capabilities will need to be evaluated for effectiveness and availability (e.g., as suggested using the fire scenario concept tree, code compliance evaluations, or performance-based egress modeling) to assess the consequence level at each point in the chronology.

10.2 Qualitative Consequence Analysis

The "consequences" term in the risk equation involves determining the potential impacts of a hazard event. Consequences can be characterized under the various classifications, typically including the following:

- Life safety.
- Property protection.
- Environmental impact.
- Business continuity.
- A combination of the above.

10.2.1 Consequence Assessment

To determine the potential for loss and associated levels of unacceptable impact from undesirable fires, consequences are typically measured in terms of health and safety impacts on people (life safety consequences), loss of property (impact on the property), business interruption costs (impact on business), or environmental damage. These consequences can be direct (e.g., the property is damaged) or indirect (e.g., the company is out of business for several days). They can be objective (e.g., replacement cost in monetary units) or subjective (e.g., pain and suffering effects of injury, utility measure of damage). The appropriate outcome should be selected to assess the risk that is dimensionally consistent with a risk acceptance criterion (as defined in Chap. 7).

It is noted that consequence analysis is more complicated than hazard evaluation. It may not always be clear how and to what extent something is valued, and the loss should be characterized. For example, in valuing life safety consequences, an application may consider only injuries and loss of life to an individual. However, there are also such factors as reduced quality of life, pain and suffering, rehabilitation after a fire-induced injury, the inability to continue to work, and the impact on family relationships.

For property protection, it may not always be clear to the interested and affected parties where, how, and how much damage may occur. Factors such as smoke and water damage should be considered, in addition to thermal damage. Demolition, environmental restoration, and rebuilding to a new code or standard can add complexity to the replacement cost calculation.

The issues can get even more complex for assessing potential business continuity impacts and damage to historically important buildings or contents. There are long-term issues, such as loss of image and market share, in addition to the short-term monetary losses associated with downtime.

Sometimes, consequences are estimated in terms of monetary values. However, the valuation in terms of monetary worth can become challenging in some cases. This is especially true for life safety, where identifying a value for human life can be complex and controversial.

In the hot work example discussed in this chapter, the event tree structure may suggest increasing levels of consequences as suppression attempts fail. Each of these consequence levels may need to consider life safety (injuries or fatalities), property protection, environmental impact, and business continuity, which depending on the source of ignition, may be generated even for non-propagating fires. Specifically, the following consequence levels are assigned, as depicted in Fig. 10.8.

- *Consequence level 1 (Fire Scenario 1):* A hot work fire that is controlled by fire prevention measures and does not propagate to nearby combustibles has negligible consequences.
- *Consequence level 2 (Fire Scenario 2):* The fire watch successfully detects and promptly controls the fire using a fire extinguisher. Such a rapid response is expected to have negligible to marginal consequences.
- *Consequence level 3 (Fire Scenario 3):* A fire that grows to an intensity where the fire watch is no longer capable of controlling it and, at the same time, capable of activating a sprinkler system may produce consequences larger than marginal. This will depend on the nature of the facility, the losses associated with the ignition source, and nearby combustible and life safety considerations.
- *Consequence levels 4 and 5 (Fire Scenario 4 and 5)* are associated with sprinkler failures and rely on the fire brigade or fire department to control the fire before flashover conditions. A fire resulting in flashover conditions is often associated with consequences ranging from major to catastrophic.

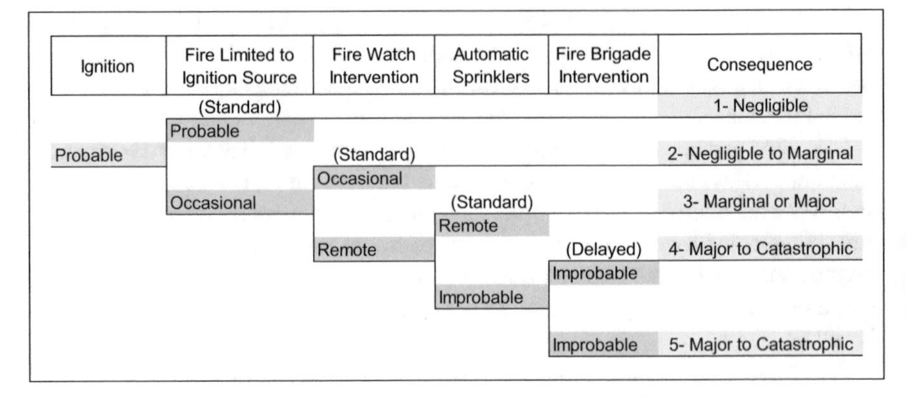

Fig. 10.8 Event tree for hot work example with assessment for consequences

With these consequence assignments, the fire risk evaluation can now proceed.

10.3 Risk Estimation

Once consequence levels are assigned to each fire scenario and their potential outcomes, the risk matrix is used to determine the level of risk associated with each scenario. The process consists of identifying the risk level specified in the matrix corresponding to the scenario frequency and consequence level assigned. Recall that Chap. 2 defined the concept of "scenario frequency," which is different from the ignition or initiating event frequency in the event tree. The term scenario frequency refers to the likelihood of the event tree sequence starting with ignition and including the effect of the conditional probabilities (as qualitatively assessed) characterizing the fire protection features.

The hot work example discussed in this chapter has five possible outcomes. Each outcome is treated as an individual scenario. That is, each has an ignition frequency already assessed as "probable," a scenario frequency characterized by fire protection capabilities evaluated as "standard" and corresponding consequences. To do so, the risk evaluation process assesses each of these five scenarios using the qualitative risk matrix in Fig. 10.9.

Continuing with the conceptual example associated with a hot work fire that has been discussed in this chapter, each of the resulting fire scenarios is qualitatively assessed.

Fire Scenario 1 is associated with negligible consequences. The ignition frequency is "probable." The likelihood of success in limiting the fire to the ignition source is "standard." The scenario frequency remains as "probable" since the success branch of the tree representing fire limited to the ignition source should be close to 1.0 as it was classified as "standard." It should be noted that because the

	Consequence				
Frequency	Negligible	Marginal	Major	Critical	Catastrophic
Frequent	Acceptable	Further Evaluation	Not Acceptable	Not Acceptable	Not Acceptable
Probable	Acceptable	Further Evaluation	Not Acceptable	Not Acceptable	Not Acceptable
Occasional	Acceptable	Acceptable	Further Evaluation	Not Acceptable	Not Acceptable
Remote	Acceptable	Acceptable	Acceptable	Further Evaluation	Further Evaluation
Improbable	Acceptable	Acceptable	Acceptable	Acceptable	Further Evaluation
Incredible	Acceptable	Acceptable	Acceptable	Acceptable	Acceptable

Fig. 10.9 Qualitative risk matrix

consequences are "negligible," it is academic as to what the combined likelihood is since there is an acceptable risk for all likelihood combinations, and no further evaluation is necessary.

Fire Scenario 2 is associated with negligible to marginal consequences. At the same time, the scenario frequency consists of a "probable" ignition followed by a failure of fire prevention measures and a successful fire watch intervention. In this example, these capabilities have been evaluated as "standard," which has the practical effect of reducing the scenario frequency and limiting the consequences to marginal levels. Therefore, Scenario 2 results in an occasional frequency (i.e., lowered at least a classification due to the credit of fire prevention measures and the fire watch capability) with marginal consequences, which is also acceptable. It is noted that fire prevention measures are explicitly credited in the analysis and should be monitored throughout the facility's operation.

Fire Scenario 3 is associated with the successful operation of automatic sprinklers limiting the consequences to "marginal" or "major," depending on the specific characteristics of the facility. Recall that this system has been classified as "standard" in the FSCT evaluation in terms of the effectiveness of the system (i.e., system design and installation to address the identified hazards) and its availability (i.e., the percentage of time the system is available to operate on demand, that is, the system is not out of service). Sprinklers are often highly effective and reliable, given they are designed for the identified hazards and the requirements for inspection and testing. In addition, they are credited in this scenario to prevent flashover conditions, which are conditions far exceeding those required for sprinkler activation. This suggests a further lowering of the scenario frequency level to "remote." The risk matrix suggests acceptable levels of risk for this configuration.

Fire Scenarios 4 and 5 are associated with major to catastrophic consequences. These scenarios result from the failure of the automatic sprinklers and the reliance of the fire brigade or fire department. Since the response time may not be early enough to prevent flashover, the scenario frequency is not further reduced from "remote." The risk matrix suggests further evaluation for these scenarios as the "remote" nature of the event, given the fire watch and automatic sprinklers already provide most of the risk reduction. Depending on the level of consequences, additional fire protection features may be necessary. At the same time, it may be practical to accept the residual risk as it is small and the existing fire protection features included in the analysis will be required to be maintained and monitored. Table 10.1 summarizes the risk evaluation results.

The risk evaluation summarized in Table 10.1 provides the following insights:

1. Importance of automatic sprinklers: This system has the practical effect of lowering the scenario frequency to a "remote" classification, which suggests that the risk is acceptable depending on the specific characteristics of the scenario. For example,

   (a) If consequences are low and based on the loss associated with the postulated scenario, the resulting risk may be tolerable or acceptable.
   (b) If the consequences are high and based on the loss associated with the postulated scenario. In that case, further evaluation may be necessary to ensure that there is enough margin in the results, that defense-in-depth strategies are available, etc.

2. If an automatic sprinkler system is unavailable or deemed ineffective, the scenario frequency may not be as low as "remote." This suggests the need to improve and monitor fire prevention strategies, the fire watch capability, and the response time of the fire brigade if these capabilities can control the identified fire hazards.
3. The concept of availability characterizing the fire protection capabilities in the analysis should be used to evaluate the level of scenario frequency reduction

Table 10.1 Summary of risk estimation results for hot work example

Scenario	Ignition frequency (initiating event)	Scenario frequency	Consequences	Risk evaluation	Comment
1	Probable	Probable	Negligible	Acceptable	Fire prevention measures prevents propagation
2	Probable	Occasional	Marginal	Acceptable	Fire prevention and hot work fire watch credited in the analysis
3	Probable	Remote	Marginal to major	Acceptable	Automatic sprinklers are credited in the analysis
4	Probable	Remote	Major to catastrophic	Further evaluation may be necessary	Fire brigade or fire department credited for suppression
5	Probable	Remote	Major to catastrophic	Further evaluation may be necessary	Fire brigade or fire department credited for suppression

assigned to the scenario. For example, a scenario frequency reduction from occasional to remote may not be justified if the sprinkler system is routinely out of service.

Although this chapter described the process for systematic qualitative risk estimation for a single ignition associated with hot work hazard developing into five fire scenarios, a full fire risk assessment should include similar evaluations for all the identified scenarios within the scope of the analysis. Each scenario will be characterized by its risk contribution to the overall risk of the facility. Chapter 11 provides additional details about risk quantification and evaluation.

References

1. SFPE, *Handbook of Fire Protection Engineering*, 5th edn. (SFPE, Gaithersburg, 2016)
2. W. Phillips, R. Fahy, Chapter 80, Computer simulation for fire risk analysis, in *SFPE Handbook of Fire Protection Engineering*, 5th edn., (SFPE, Gaithersburg, 2016)
3. NFPA, *NFPA 550, Guide to the Fire Safety Concepts Tree* (NFPA, Quincy, MA, 2022)
4. U.S. Nuclear Regulatory Commission/EPRI, *EPRI/NRC-RES, Fire Human Reliability Analysis Guidelines. NUREG-1921* (U.S. Nuclear Regulatory Commission/EPRI, Washington, DC/Palo Alto, 2012)

Chapter 11
Quantitative Fire Risk Estimation

This chapter builds on the qualitative risk estimation process, and the example described in Chap. 10 provides guidance on quantitative risk estimation. Quantitative risk estimation refers to the process of determining ignition frequencies, conditional probabilities, and consequences in numerical terms to calculate a risk value for each scenario contributing to fire risk in a facility.

This chapter does not aim to provide a detailed description of the available frequency and consequence prediction approaches but rather an overview of selected examples that may be employed in a fire risk assessment, considering the nature and detail of information typically collected from fire incidents.

The output of the process described in this chapter generally consists of a table or list of fire scenarios with the corresponding frequency and consequence and their related risk estimates. Depending on the objectives, the risk estimates for an individual scenario, groups of scenarios, or the entire "risk profile" for a system/facility/building will then be evaluated for acceptance.

11.1 Quantitative Estimation of Fire Risk

A quantitative assessment evaluates risk based on numerical estimates of scenario frequencies (i.e., the multiplication of ignition frequencies and conditional probabilities in the sequence of events characterizing a fire scenario) versus the potential consequences of fire events postulated in the analysis. Both the frequency and the consequence of the fire scenario are evaluated in values consistent with the way risk is evaluated (i.e., values that have the same units). The quantitative nature of the analysis requires identifying and categorizing the factors affecting fire scenario frequencies and the corresponding consequences and expressing them numerically.

SFPE Guide to Fire Risk Assessment, The Society of Fire Protection Engineers Series, https://doi.org/10.1007/978-3-031-17700-2_11

In general, for each scenario defined in the risk assessment, the analyst should:

- Quantify an ignition frequency
- Determine the fire protection features' ability to limit the consequences associated with the potential outcomes of the fire scenario through quantifying conditional probabilities
- Quantitatively assess the consequence level for each potential outcome of the fire scenario

Continuing with an event tree to represent a fire scenario as presented in Chap. 9, the ignition frequency and the conditional probabilities are multiplied to obtain each sequence's frequency. At the same time, each sequence in the scenario progression event tree is associated with a consequence level. The frequency and consequences are then combined (e.g., multiplied) to obtain a numerical description expressing the risk that can be used for ranking scenarios and decision-making. Finally, the total fire risk for the facility can be quantified as the sum of all scenario's risk contributions.

As in qualitative assessments, the quantitative assessment of frequency, fire protection capabilities, and resulting consequences are based on the review of fire event records, engineering judgment, or analytical modeling. Chapter 10 describes these methods.

11.2 Quantitative Frequency Analysis

The ignition frequency characterizes the likelihood of fire initiation, representing the starting point of a fire scenario. Quantitatively, the frequency is the ratio of the number of times an event occurs in a time period and is mathematically represented as shown in Eq. 11.1.

$$\lambda_i = \frac{n}{T} \tag{11.1}$$

where:

λ = the frequency of ignition in the fire scenario i.
n = the number of events
T = the time period

The time period is the relevant period of time selected for data collection. This estimate assumes the frequency is constant throughout time, which is a reasonable and simplifying assumption often made in fire risk assessment.

Determining ignition frequencies for a fire risk assessment usually requires a well-planned effort for collecting, classifying, and analyzing data for the application. This effort should not be underestimated, as determining the applicability or relevance of the available data can be a significant effort. In general, the scope of the data collection process is governed by:

- *Time periods:* This refers to the period of time in which relevant/applicable data is available.
- *Technology:* Due to continuously evolving technologies, new products in the markets may introduce new fire event patterns influencing the scope of the data collection process.
- *Applications:* The data collection process may be limited to specific industries or facilities to ensure the applicability of the resulting risk. Some industries, for example, have specific maintenance and operational requirements that would suggest that the fire incident would produce a particular risk profile.
- *Fire safety regulations:* New or updated fire safety regulations may alter the pattern of fire events observed during different time periods or geographical areas.
- *Other societal trends or patterns:* Societal patterns may also influence the selected time period for data collection.

Once relevant data has been collected, it needs to be analyzed and classified for quantification purposes. The analysis and classification process is intended to ensure that the ignition frequency values calculated are independent of the ability to include other conditional probabilities in the risk equation. For example, if the risk equation is defined, the frequency term represents ignition, the data selected should contain all the events where ignition "if left alone" would have triggered a fire event. This would allow the inclusion of conditional probabilities representing the ability to detect and suppress the fire in the quantification process. These fire protection features were not considered part of the criteria for selecting data used for calculating the ignition frequency. If, on the other hand, the data selected for the calculation of the ignition frequency would have excluded events where fires never grew due to the activation of automatic systems or human intervention, the inclusion of conditional probabilities in the risk equation would inappropriately "double count" the effects of fire protection features in the resulting risk estimates.

This expands on the discussion related to the scenario associated with hot work in a facility protected by automatic sprinklers described in Chap. 10 and depicted in Fig. 10.1. Also recall that procedures are in place requiring a fire watch to be posted while hot work activities are ongoing. Firstly, the analyst should research the general trends of hot work fires. A cursory review of hot work fire events suggests the following for manufacturing facilities (it is noted that these are approximate values used for the example and should not be cited for specific applications):

- Number of manufacturing facilities (including all types/sizes): Approximately 300,000
- Number of reported fires responded by fire department per year in manufacturing or industrial facilities: 37,000
- Percentage of fires due to hot work: 20%

These generic values, often available from national fire statistics or other studies, should be carefully considered as they may not reflect ignition frequency as represented in the fire risk assessment. The number of fires responded to by a fire department may only suggest those events in which the fire grew to intensities requiring a fire department response. Alternatively, the fire department may have also responded to an alarm, but the fire watch had easily controlled the fire. Using these generic values results in an estimated frequency of:

$$\lambda = \frac{(37{,}000)(0.2)}{300{,}000} = 2.47E - 2\,\text{fires / year}$$

Where 300,000 is the number of facility-years of operation and the multiplication in the numerator is the number of hot work fires. At the same time, the analysts should review the specific history of fires in the facility within the scope of the study. For this example, it is assumed that there has been one fire event report due to hot work activities in the facility since the start of operations 12 years ago. This results in:

$$\lambda_i = \frac{n}{T} = \frac{1}{12} = 0.08\,\text{fires / year}$$

It is important to note that basing the frequency only on the facility-specific data may not capture fire patterns in similar facilities. Situations such as the facility under study are still in the design phase or have not experienced fires. These are some of the reasons for which a broader perspective on ignition frequencies is necessary. The information available related to fire events due to hot work experience with the qualitative assessment is in the range of "probable" to "occasional," as discussed in the previous example, suggesting that these are events expected to happen in the order of one every 10 years (i.e., recorded sometime throughout the lifetime of a facility). It is noted that the frequency characterization is limited to the ignition event, regardless of the consequences these events have generated in the past. In that way, the frequency characterization is independent of fire protection features or mitigative strategies explicitly credited later in the analysis to reduce the potential consequences of the scenario. Figure 11.1 below depicts the quantitative

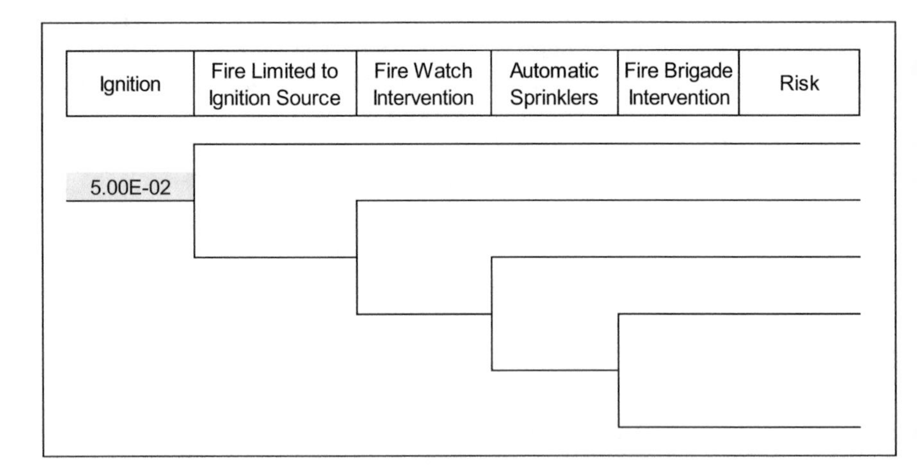

Fig. 11.1 Event tree for hot work example with quantitative ignition frequency assessment

frequency assessment for this example. Description of statistical techniques to update generic frequency values with facility-specific data (e.g., Bayesian updates) is out of the scope of this guide. For this example, the frequency information described earlier is used to assess a range of potential values for the ignition frequency. As such, the value of 0.05 fires/years is selected for the analysis as approximately the average of both values. This value is bounded by the generic and facility-specific data. It should later be the subject of a sensitivity and uncertainty analysis to determine its influence on the decision-making process.

11.2.1 Quantitative Assessment of Conditional Probabilities

Recall that conditional probabilities are factors representing specific elements of a fire scenario, such as fire protection features. As described in the previous chapter, this specifically refers to the fire's ability to propagate, the effectiveness, reliability, and availability of procedures or systems to be performed as designed in the context of the postulated fire scenarios. Several methods are available to systematically quantify conditional probabilities capturing the systems' effectiveness, reliability, and availability. This section conceptually describes some of those methods.

Conditional Probabilities for Fixed Ignition Sources Recall that the ignition frequency assigned to a fixed ignition source should represent that source individually (i.e., a frequency value per ignition source) to avoid "overcounting" the risk contribution when the total risk of the facility is evaluated by adding the contribution of each ignition source. If an ignition frequency per individual source is available, no conditional probability is necessary. If instead, frequency representing the likelihood of ignition of a type of ignition source in a facility is available, such value may need to be apportioned to each of the ignition sources of that type identified in the facility.

Conditional Probabilities for Transient Fire Locations Within Multiple Areas in a Facility Since transient ignition frequencies are not associated with a specific location within a facility, a conditional probability for distributing the frequency within the different areas of a facility may be necessary. This conditional probability serves as an apportioning factor to distribute the total frequency throughout the different areas. The apportioning factor can be assigned based on the facility's maintenance, storage, and operational characteristics. Factors such as storage of combustibles or flammable liquids, occupancy, and the level of maintenance activities in the different areas of the facility could be used as weighting factors for a realistic apportioning of transient ignition frequencies. For example, a much lower transient frequency may be assigned to an area of the facility where hot work is never performed. To develop the apportioning factor, the analysts may classify the different areas in the facility according to levels characterizing these factors (e.g., low, medium, high maintenance activities). An approach [1] for developing these apportioning factors is based on:

- *Level of occupancy:* This can be associated with the number of people or foot traffic associated with the different areas within the facility. Higher occupancy or foot traffic may suggest a higher frequency of transient fires when compared to empty areas within the facility.
- *Level of hot work and/or maintenance:* Areas within the facility where relatively larger numbers of maintenance or hot work activities are performed may be assigned higher transient frequency values than those where such activities are rare.
- *Level of storage:* This refers to the storage of transient ignition sources, combustibles, or flammable liquids within the different areas in the facility. A higher level of storage may result in a higher likelihood of transient fires when compared to areas with no storage. This apportioning factor can explicitly capture combustible control procedures assigned to specific areas in the facility in the fire risk assessment.

To develop apportioning factors, the level of occupancy, maintenance/hot work, and storage are determined individually for each area within the facility. As a practical approach, this ranking is shown as follows:

- Very low, with a value of 0.1
- Low, with a value of 1.0
- Medium, with a value of 3
- High, with a value of 10
- Very high, with a value of 50

The numerical values suggested above are not linear to capture realistic operational differences between areas in the facility (e.g., differences between a manufacturing floor area and an office space within a facility). Ranking values of zero are not recommended (unless physically prevented or the hazard is eliminated by design) to avoid setting a probability of a transient fire in an area to zero. Furthermore, as a practical approach, "Medium" ranking values may be treated as default, increasing and decreasing the areas systematically from a default value as the operational characteristics of each area are evaluated.

To illustrate the apportioning process, consider a facility with three areas: Area A, Area B, and Area C. Each area is ranked for the level of occupancy maintenance/hot work and storage, as summarized in Table 11.1.

Using the values summarized in Table 11.1, the apportioning factor for each area is calculated as:

$$\text{Area A}: \frac{11.1}{23+53.1+12} = \frac{11.1}{88.1} = 0.126$$

$$\text{Area B}: \frac{7}{23+53.1+12} = \frac{7}{88.1} = 0.079$$

Table 11.1 Example of ranking for occupancy maintenance/hot work and storage in three areas within a facility

Area	Occupancy	Maintenance/hot work	Storage	Total per area
Area A	High, 10	Very low, 0.1	Low, 1	11.1
Area B	Medium, 3	Medium, 3	Low, 1	7
Area C	High, 10	Very high, 50	High, 10	70
Total:	**23**	**53.1**	**12**	

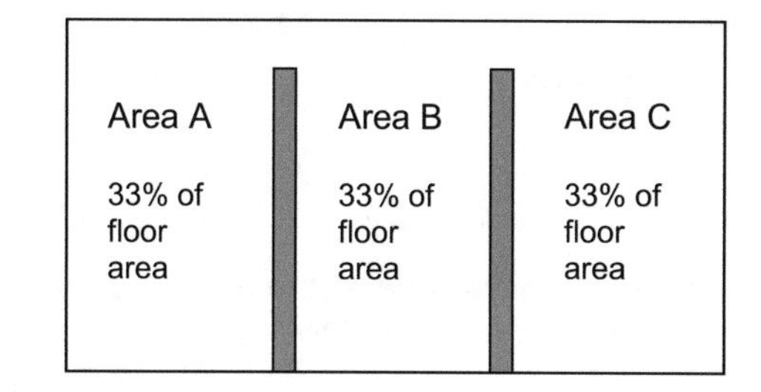

Fig. 11.2 Pictorial representation (top view) of the concept of floor area ratio to specify the location of a transient fire scenario in the quantification process

$$\text{Area C}: \frac{70}{23+53.1+12} = \frac{70}{88.1} = 0.0.794$$

Conditional Probabilities for Transient Fire Location Within an Area As discussed earlier, a transient ignition source does not have a specific (i.e., fixed) location in the facility as it could occur anywhere where it is "physically" possible. Therefore, the probability of a transient fire in a specific location needs to be captured in the analysis. The concept of "floor area ratio" is often used to represent this conditional probability. The "floor area ratio" represents the fraction of the applicable floor area in the facility where the transient fire scenario has been defined. Using this concept, the transient ignition frequency that has been determined from data can be apportioned throughout the facility in a systematic approach that prevents the "overcount" of frequencies when multiple scenarios are postulated.

In the hot work example, consider that it has been determined that only three hot work scenarios will be postulated in the assessment. Each of these scenarios has a different location so that all possible locations where hot work could be performed are captured in the analysis. Consequently, the floor area associated with each scenario is used to determine the fraction (i.e., the conditional probability) of the total applicable floor area corresponding to the scenario. For this conceptual example, a facility is assumed with areas A, B, and C having identical floor areas. Therefore, a 0.33 probability is assigned to each, as illustrated in Fig. 11.2.

The probability of 0.33 is included in the event tree for each scenario as part of the fire ignition frequency, as illustrated in Fig. 11.3.

Conditional Probabilities for Fire Severity The second event in the sequence event tree presented in Fig. 11.1, "fire limited to the ignition source," is the first conditional probability in the analysis. It represents the fire potential to propagate

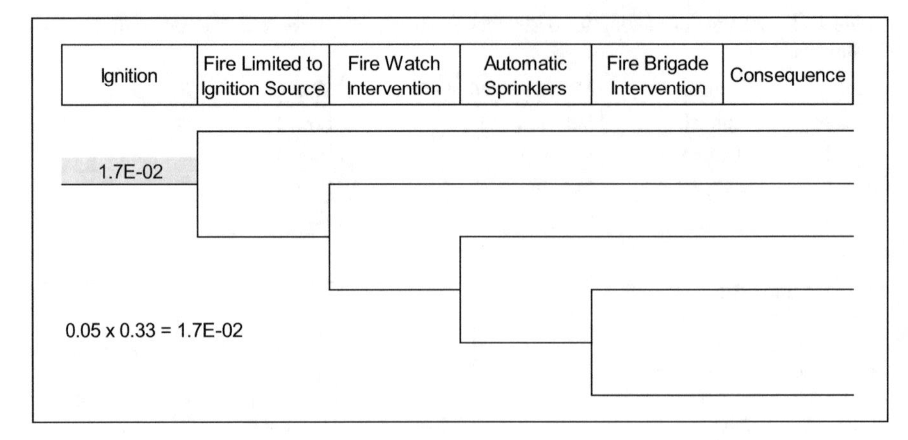

Fig. 11.3 Event tree for hot work example with assessment for conditional probability of hot work fire occurring in a specific location

outside the ignition source. This probability is used because an ignition event, as characterized in the frequency term, does not warrant an assumption that the fire will become a challenging fire capable of damaging property or impacting life safety.

The concept of fire large enough to propagate has been represented in fire risk assessments using conditional probabilities in two forms:

- Using "split fractions," in which the probability represents the likelihood of propagation given ignition. This probability can be determined from the available fire events data independent of the ignition frequency calculation. The concept of independence ensures that the fire's ability to propagate is identified regardless of the event's outcome, which may have been "altered" by the intervention of fire protection features. It is recommended that the fire events review be performed systematically by establishing specific criteria for classification so that the analysis can be reviewed and updated as necessary. For example, the available fire events data can be analyzed and classified based on the ability of the fire to propagate after ignition. This would result in a subset of data classified based on fire propagation that can then be used to estimate the fraction of fire events that are expected to propagate. This approach has the disadvantage of not capturing the specific configuration of a scenario as a "generic" split fraction is based on the configurations captured in the available data.
- The commercial nuclear industry has developed an approach for calculating the conditional probability of fire severity that addresses the disadvantage discussed earlier on using split fractions. In this approach, the intensity is characterized by a probability distribution representing the uncertainty associated with the fire size. In other words, the identified ignition source could generate different fire intensities (i.e., heat release rate profiles) in a fire event. The propagation probability is then associated with the fraction of the fires with intensities equal to or higher than those required for propagation.

Applying the concept of fire severity to the hot work example, consider the probability of an ignition event with sufficient intensity to propagate to nearby combustibles. In the context of hot work, this consists of sparks or ignited welding slags in contact with a nearby combustible identified as the first item ignited. This first item ignited is then assigned a probability distribution for its heat release rate. That is, a heat release rate probability distribution is assigned to the first item ignited, which captures the uncertainty associated with its heat release rate. Once the distribution is available, the severity factor is the area under the distribution curve to the right of the heat release rate value that will generate damage or propagation outside the ignition source. This definition of fire severity captures:

- The flammability properties of the ignition source, as the heat release rate is characterized by a probability distribution.
- The geometry characteristics of the fire scenario, such as the distance from the ignition source to propagating combustibles, are explicitly treated in the analysis.

For items nearby the ignition source, the severity factor may be near 1.0 since low heat release rates may have the capability of generating damage outside the ignition source. In contrast, for equipment located relatively far from the ignition source, larger heat release rates may be needed for causing damage, resulting in relatively low severity factors values. This concept is represented in Fig. 11.4, where f(Q) is the probability distribution for the heat release rate. Notice that the area under the distribution to the right of the heat release rate necessary for propagation represents the probability of a fire equal to or larger than what is necessary.

Recent research funded by the commercial nuclear industry has developed probability distributions for selected items that could be postulated as ignition sources in fire scenarios [2–5]. For this example,

- A scenario in which the first item ignited can produce a fire large enough to generate the postulated consequences would result in a conditional probability of a severe fire of 1.0. That is, if ignited, the postulated range of consequences is expected.

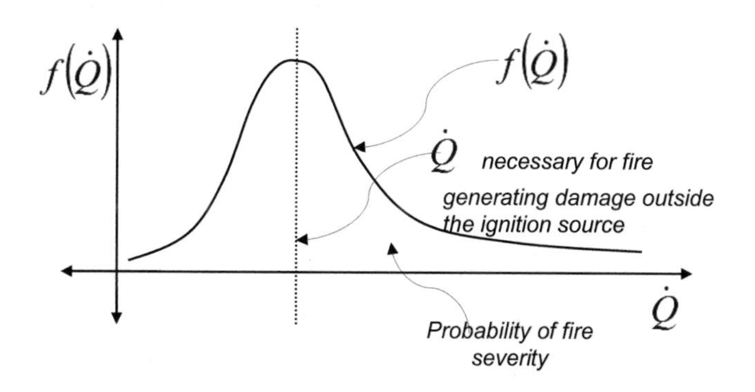

Fig. 11.4 Conceptual representation of the probability of fire severity

- A scenario in which the first item ignited is small enough not to produce propagation and negligible consequences are expected would result in a conditional probability of a severe fire close to zero.
- A third possible configuration consists of a first item ignited, propagating, and growing to conditions capable of generating the range of postulated consequences.

Example

To illustrate the third point listed above, consider a room where hot work activities are performed. The room has storage of ordinary combustibles within a caged area (e.g., chain link fenced area) as illustrated in Fig. 11.5. A transient fire is postulated outside the caged area. The transient fire is characterized by a gamma probability distribution with parameters $\alpha = 0.271$ and $\beta = 141$. This is the distribution for transient fires described in Table 4.1 in NUREG-2233 [3] based on fire testing documented in NUREG-2232 [2]. The minimum distance between the fire and the combustibles is 0.6 m. Based on material flammability properties, it is assessed that the ordinary combustibles may ignite if exposed over time to a heat flux of 15 kW/m^2. A fire modeling analyst determines that a 230 kW fire can generate flame radiation levels of 15 kW/m^2 0.6 m away. Therefore, the fraction of fires that can produce 15 kW/m^2 or more is the area under the curve of the gamma probability distribution to the right of 230 kW, or what is equivalent to the probability of $Q > 230$ kW where Q is the random variable for the heat release rate. This can be solved numerically with Microsoft Excel using the gamma probability distribution function as:

$$= 1 - \text{GAMMADIST}(230, 0.27, 141, \text{TRUE}) = 0.03,$$

which solves for the area under the distribution to the right of 230 kW.

The resulting value of 0.03 is interpreted as the conditional probability of a transient fire igniting secondary combustibles (i.e., propagating outside of the first item ignited). In this example, the chain link fence is explicitly included in the analysis as a fire protection feature separating potential fires from the combustibles. Therefore, this is a configuration that should be monitored as the resulting probability could be as high as 1.0 if the barrier is not considered as the fence protects the

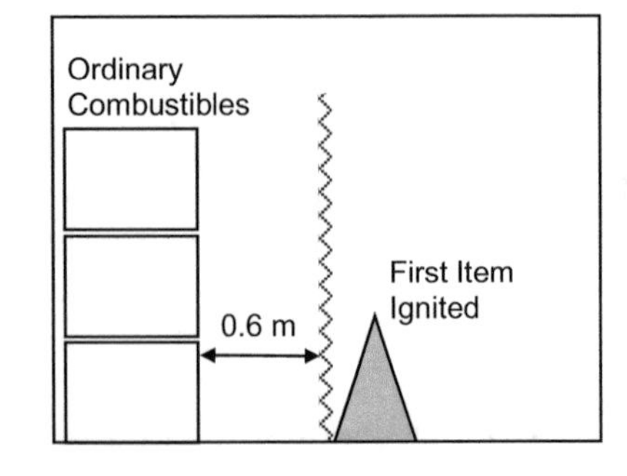

Fig. 11.5 Conceptual representation of the scenario configuration and the determination of conditional probability for fire propagation

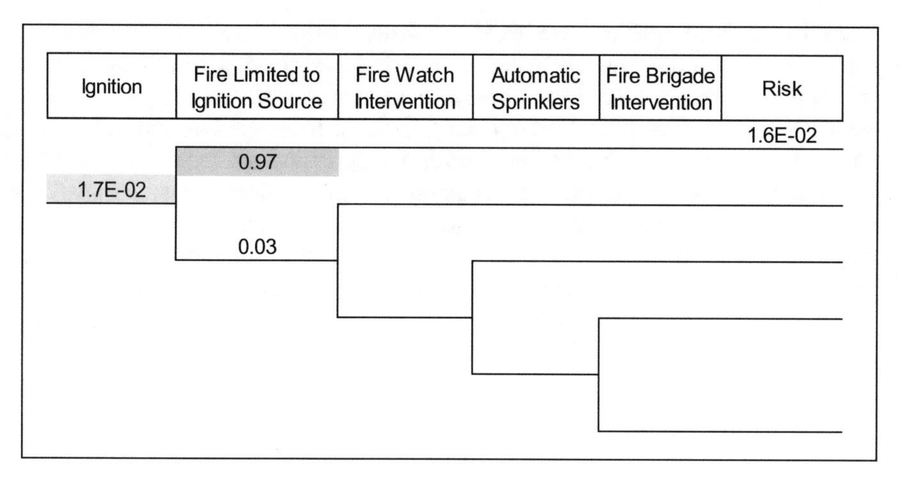

Fig. 11.6 Event tree for hot work example with assessment for conditional probability for fire severity (propagation outside the ignition source)

ordinary combustibles from ignition sources that could potentially be placed next to them. Hot work activities near the ordinary combustibles without protection would result in higher risk estimates. Figure 11.6 presents the scenario event tree with the conditional probability for fire severity.

Conditional Probabilities for Failing Fire Prevention Practices The application of this conditional probability depends on how the ignition frequency data was developed. Likely, available fire events data may already reflect the impact of fire prevention practices in reducing the likelihood of ignition. If the ignition frequency data was developed by selecting fire events where the combustible control procedures had failed (i.e., fire events where fire prevention procedures successfully prevented propagation after ignition), then the conditional probability should not be included in the analysis as the frequency value already captures the effects of combustible controls. On the other hand, if fire events selected for the frequency calculation include those in which the fire prevention procedures successfully prevented propagation, then a conditional probability is necessary to capture this capability in the analysis explicitly. The latter is usually not likely as fire events data is often collected from similar industries or facilities governed by similar fire prevention practices.

It is noted that failure of combustible controls and other similar fire prevention procedures can be considered pre-initiators (latent) human failure events. They happen "before" ignition. Furthermore, the scope of the assessment of this conditional probability does not include the failure related to fire systems, barriers, or other elements of programs outside combustible controls. Undetected pre-initiator human failures such as improperly restoring fire suppression equipment after test, compromising fire barriers are treated in this guide separately from other conditional probabilities.

Conditional Probabilities for Fire Watch Intervention Fire watch intervention can be expressed quantitatively using a probabilistic approach based on the data. In this approach, fire events data is analyzed to characterize the time for manual response to a fire event. A subset of events, those with enough information associated with fire response, fire control, and fire suppression are used to calculate a probability of failure to suppress as a function of time. Generally, data for analysis consists of reported fire durations as available in the fire event descriptions. These durations are treated as being generated by an underlying probabilistic model. The final output of interest would be a suppression curve, which gives the probability that a fire lasts longer than a specified time (i.e., the time specified in the fire scenario event tree). The probability of non-suppression is shown in Eq. 11.2.

$$\Pr(T > t) = \exp\left(-\int_0^t \lambda(s)\,ds\right) \tag{11.2}$$

where:

$T = T$ is the random variable describing when the fire is suppressed (i.e., the time to suppression)
$\lambda(s)$ is the rate at which the fire is suppressed (possibly time-dependent)
t = time

The simplest probability distribution for T is the exponential distribution, whose probability density function is shown in Eq. 11.3.

$$f(t) = \lambda e^{-\lambda t} \tag{11.3}$$

where:

$f(t) = a$ function of the parameters of the probabilistic distribution chosen for T.
This exponential distribution is often used as its parameter λ represents a constant rate, which fits this application as it can be directly calculated from data. That is, in this model, λ is estimated directly from the data and assumed constant (i.e., not a function of time), as shown in Eq. 11.4.

$$\Pr(T > t) = e^{-\lambda t} \tag{11.4}$$

The non-suppression probability is calculated using Eq. 11.4, usually selecting t as the time identified in the fire scenario event tree.

The latest research in the commercial nuclear industry (see [6] Table 5.1) for the response time from a hot work fire watch suggests an average response time of 9.3 min. This corresponds to a rate for which fires are suppressed of $\lambda = 1/9.3$ min = 0.107. This rate can be used in applications where:

- Procedures governing fire watch duties are available.
- Fire watches are trained, including training in fire suppression using fire extinguishers.
- The fire watch acts on a fire that can be suppressed with a fire extinguisher (i.e., a fire that starts small propagates relatively slowly due to ignition from welding spark or ignited slag).

The mean rate of $\lambda = 0.107$ fires per minute is reported in the same reference as having the fifth percentile of 0.084 fires per minute, 50th percentile of 0.107, and 95th percentile of 0.133 fires per minute. This suggests that an approximately symmetrical distribution represents the uncertainty associated with this parameter. The fifth percentile rate can be used for a conservative approach, representing the fire watch's longest response time.

Example
In the hot work example in this section, assume the event "fire watch intervention" is postulated to occur within 5 min. Using the conservative suppression rate of $\lambda = 0.084$, the resulting non-suppression probability for this time step is:

$$\Pr\left(T > t\right) = e^{-lt} = e^{-0.084 \cdot 5} = 0.65$$

Conditional Probabilities for Failure to Operate Manual Systems Recall from Chap. 10 that this probability can be calculated using human reliability techniques, and it often includes a qualitative and quantitative assessment. The qualitative assessment is limited to determining if the action is feasible. The quantitative assessment evaluates how reliable a feasible action is by assigning a human error probability.

For actions determined to be feasible, which means that they are likely to be completed in time successfully, a human error probability can be calculated to support the risk quantification process. Recent research [7] in the commercial nuclear industry has developed detailed methods for quantifying human error probabilities. It is noted that although the calculation of human error probabilities can be complex, particularly for those operation actions involving the execution of lengthy procedures in relatively short periods of time, the process can be significantly simplified for relatively simple actions such as starting a manual fire suppression system. The simplification consists of assigning "screening" values intended to be conservative to the human error probabilities. Specifically,

- Assign a value of 1.0 (i.e., guaranteed failure of the action) if there is no margin in the time available to complete the action or the action is not feasible. A value of 1.0 may also be assigned if the time margin is minimal (i.e., there is no time for correcting any error made in the execution of the action).
- A screening value of 0.1 can be assigned if there is a margin in the time available to complete the action. Values between 0.1 and 1 could also be assigned for lower margins in time. The concept of margin depends on the specifics of the scenario.

In the context of activation of manual systems, clear indicators or cues related to identifying a fire event (e.g., fire detection alarm) would reduce the need for a large margin in time, assuming personnel is trained on how to proceed upon receiving the alarm. In addition, the available margin is enough to correct or repeat any error made in the execution. To assign this value, the fire-generated conditions (e.g., the presence of smoke) should not affect the ability to complete the action.

- For actions where there is a margin in time, procedures are available; personnel have received training on the procedures, the equipment necessary to perform the action is readily available:
  - Assign a value ranging from 2E-3 to 0.01 for low complexity actions (i.e., the action is relatively easy to execute). A value of 0.01 would be considered conservative in the context of human reliability analysis. Lower values within that range could be assigned based on the qualitative assessment.
  - Assign a value ranging from 0.01 to 0.05 for complex actions. A value of 0.05 would be considered conservative in the context of human reliability analysis. Lower values within that range could be assigned based on the qualitative assessment.

These screening values could be further lowered using detailed human reliability analysis techniques. The description of such models and techniques is outside of the scope of this guide.

Finally, the resulting conditional probability needs to include the contribution from the hardware failure and the human failure. In practice, these two values are added together.

Conditional Probabilities for Detection and Suppression Following Prompt Response Detection and suppression can be interpreted as a "non-suppression probability" representing the likelihood that the fire is not suppressed before the postulated damage during the scenario time interval. These probabilities are often conditional upon previous scenarios. As such, a non-suppression probability needs to be calculated for each point in time defined in the scenario progression event tree conditional on the ability to suppress at the previous time.

An "event tree" model is a convenient tool for determining the non-suppression probabilities as a function of time as it captures the chronology of fire protection features included in the analysis. It also provides the advantage of incorporating the most active detection and suppression capabilities as appropriate for each scenario often included in a fire risk assessment. An event tree that could be used to calculate non-suppression probabilities is illustrated in Fig. 11.7. It is noted that this event tree can be considered "generic" and may need to be adjusted for specific applications, as necessary.

The following sequence of detection and suppression capabilities are captured in the event tree after the "ignition" initiating event:

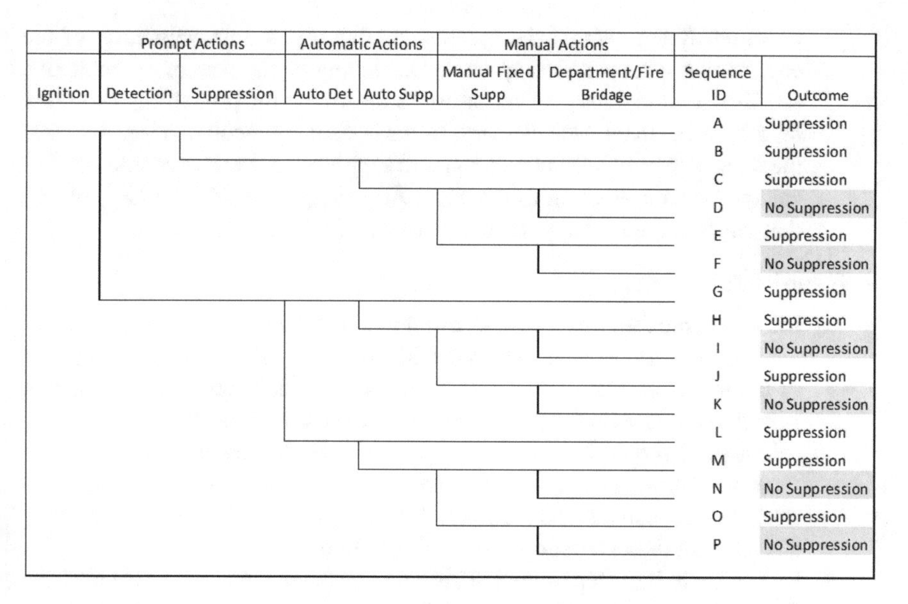

Fig. 11.7 Example detection/suppression event tree model for calculating non-suppression probabilities

1. Prompt Capabilities

   (a) Prompt Detection refers to the ability of people (e.g., facility personnel, building occupants, fire watch) present or nearby the location of fire origin to detect and alert about the fire event. Quantification of this prompt response is often represented with high reliability (i.e., a probability of failure close to zero) as humans are usually capable of identifying fire conditions in the early stages of the event.

   (b) Prompt Suppression refers to the ability of people present or nearby the location of the fire origin to suppress or control the fire in its early stages, reducing the likelihood of growth and propagation. Recommendations for quantification of prompt suppression capabilities were discussed earlier in this chapter associated with fire watch response.

2. Automatic Capabilities

   (a) Automatic Detection: The first event after ignition is the operation of the automatic detection system. This can be numerically represented by the probability of the detection system failing to operate on demand, which is often captured by the sum of the system unreliability and unavailability. Notice that even when smoke detection fails, the event tree captures the probability of suppression by automatic sprinklers (if available and the systems are independent).

   (b) Automatic Suppression: This event refers to the probability of failure on demand of the suppression system. The failure to operate on demand is often captured by the sum of the system unreliability and unavailability of the automatic fire suppression systems. Depending on the system's complexity, a fault tree model may be necessary to determine the probability of failure on demand so that all the elements of the system are accounted for. Note that the dependency between detection and suppression systems should be considered when evaluating this branch. An example would be a CO_2 system that requires the actuation of an automatic heat detection system.

3. Manual Capabilities
   (a) Manual Fixed Suppression: This event refers to the unreliability of the manual fixed fire suppression systems. Guidance on quantifying the probability of failure to activate a manual system was provided earlier in the chapter. Recall that this should include both the human error and the hardware contribution. It should be noted that any dependency between automatic and manual systems should be addressed in the model. For example, if an automatic system can also be manually actuated, a failure in the hardware may also prevent its successful automatic operation.
   (b) Fire Brigade/Fire Department: This captures the manual suppression by the fire brigade or Fire Department. The methodology described earlier to numerically characterize the fire watch response can be used for the fire brigade or fire department response. A mean suppression rate will need to be available for the response of the fire brigade or fire department for different types of fire scenarios.

The non-suppression probability results from adding branch probabilities D, F, I, K, N, and K are those failing to detect and suppress.

Each event in the tree requires consideration of time. For example, when determining if the automatic sprinkler system should be credited in the analysis, the analyst should compare the specified time in the scenario progression event tree for that specific event and the time for sprinkler activation. If the time specified in the event is shorter than the time to sprinkler activation, then the system should be modeled with a failure probability of 1.0.

Continuing with the hot work example to illustrate the use of the event tree model, consider Table 11.2, which summarizes the scenario progression event tree. If the fire is suppressed within 5 min (Scenario 1), the damage will be limited to the ignition source. If the fire burns for more than 5 min but less than 15 min before it is extinguished, the damage will be limited to the ignition source and nearby combustibles where the fire propagated. In Scenario 3, the fire burns between 15 and 25 min resulting in further consequences. Finally, there is an additional scenario in which the fire is not suppressed. This results in the worst scenario outcome (i.e., flashover).

In this example, the fire protection program provides the following capabilities for detection and suppression: Prompt detection and suppression by the fire watch, automatic smoke detection system, automatic sprinklers, and the fire brigade. To calculate a non-suppression probability as a function of time given these

Table 11.2 Example of scenario refinement based on fire suppression consideration

Ignition source	Event	Elapsed time after fire ignition (min)	Consequences
Hot work	Fire limited to ignition source	N/A	Damage limited to ignition source
Hot work	Fire watch intervention	5	Damage limited to ignition source
Hot work	Automatic sprinklers	15	Damage to ignition source and propagating combustibles
Hot work	Fire brigade intervention	25	Further fire propagation before the onset of flashover

capabilities, the information listed in Table 11.3 is necessary. The first column in the table lists different capabilities available in the program for both detection and suppression. The second column in the table indicates if the capability is included in the fire scenario analysis to calculate a non-suppression probability. The third column shows the expected time for the system to operate or the capability to be available. This timing information can be generated from fire modeling calculations or operating experience (e.g., reviewing training records or similar fire events). The last column, "Availability," refers to the combination of reliability and availability, capturing the probability of failure on demand. The values presented in this table are arbitrary numbers selected for the example.

In Table 11.3, the calculation for the failure probability for the fire watch suppression capability was described earlier in this chapter and resulted in a value of 0.65. The same approach is used for the fire brigade. Since the time for prompt detection for the fire watch is 0 (i.e., the fire is immediately detected), it is assumed that this detection action will trigger a response by the fire brigade. For this example, a suppression rate of 0.1 fire/min is assumed. Therefore, the resulting suppression failure probabilities for the fire brigade are:

- For responding within 5 min: $Pr(T > 5) = 1.0$, as the brigade is not expected to respond within 20 min.
- For responding within 10 min: $Pr(T > 15) = 1.0$, as the brigade is not expected to respond within 20 min.
- For responding within 25 min: $Pr(T > 25) = e^{-\lambda t} = e^{-0.1(25 - 20)} = 0.6$. Notice that the time available for suppression is 5 min, given a response time of 20 min.

Using the input values listed above, the event tree in Fig. 11.7 can be solved for the 5, 15, and 25 min events defined in the scenario progression event tree, as illustrated in the following three event trees. When solved for 5 min, the resulting non-suppression probability is 0.35 (See Fig. 11.8), reflecting the fire watch capability of controlling or suppressing the fire within that time period. At this time, the sprinklers have not activated, and the fire brigade has not arrived. Recall that it has been assumed that the fire watch's prompt detection triggers the fire brigade's response for this example. This is not a necessary assumption as both the trigger and response time for the fire brigade or fire department should be represented realistically for each application.

Table 11.3 Summary of inputs to the detection/suppression event tree model

Fire protection capabilities	Include in the scenario	Activation time (min)	Availability
Detection			
Automatic smoke detection	TRUE	5	0.95
Automatic heat detection	FALSE	0	0
Prompt detection (personnel, fire watch)	TRUE	0	1.0
Late/delayed personnel detection[a]	TRUE	10	0.95
Suppression			
Prompt suppression (personnel, fire watch)	TRUE	0	1 – 0.65 = 0.35
Automatic sprinklers	TRUE	12	0.98
Automatic halon	FALSE	0	0
Automatic CO_2	FALSE	0	0
Manually activated system	FALSE	0	0
Fire brigade/fire department	TRUE	20	At 5 min: 0
	TRUE	20	At 10 min:0
	TRUE	20	At 25 min: 1 – 0.95 = 0.05

[a]It is assumed that the fire will be eventually detected if automatic detection or suppression fails. In this example, the presence of a fire watch suggests a very high probability of quick detection

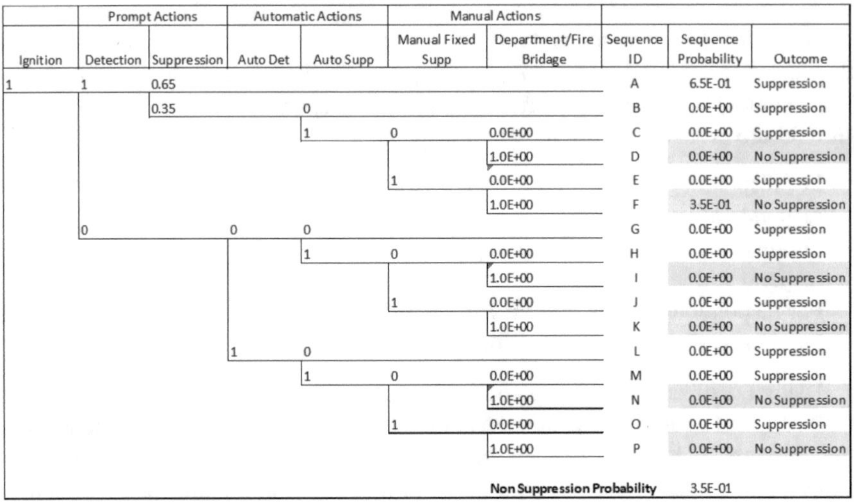

	Prompt Actions		Automatic Actions		Manual Actions				
Ignition	Detection	Suppression	Auto Det	Auto Supp	Manual Fixed Supp	Department/Fire Bridage	Sequence ID	Sequence Probability	Outcome
1	1	0.65					A	6.5E-01	Suppression
		0.35		0			B	0.0E+00	Suppression
				1	0	0.0E+00	C	0.0E+00	Suppression
						1.0E+00	D	0.0E+00	No Suppression
					1	0.0E+00	E	0.0E+00	Suppression
						1.0E+00	F	3.5E-01	No Suppression
	0		0	0			G	0.0E+00	Suppression
				1	0	0.0E+00	H	0.0E+00	Suppression
						1.0E+00	I	0.0E+00	No Suppression
					1	0.0E+00	J	0.0E+00	Suppression
						1.0E+00	K	0.0E+00	No Suppression
			1	0			L	0.0E+00	Suppression
				1	0	0.0E+00	M	0.0E+00	Suppression
						1.0E+00	N	0.0E+00	No Suppression
					1	0.0E+00	O	0.0E+00	Suppression
						1.0E+00	P	0.0E+00	No Suppression
						Non Suppression Probability		3.5E-01	

Fig. 11.8 Detection/suppression event tree solved at 5 min

	Prompt Actions		Automatic Actions		Manual Actions				
Ignition	Detection	Suppression	Auto Det	Auto Supp	Manual Fixed Supp	Department/Fire Bridage	Sequence ID	Sequence Probability	Outcome
1	1	0.65					A	6.5E-01	Suppression
		0.35		0.98			B	3.4E-01	Suppression
				0.02	0	0.0E+00	C	0.0E+00	Suppression
						1.0E+00	D	0.0E+00	No Suppression
					1	0.0E+00	E	0.0E+00	Suppression
						1.0E+00	F	7.0E-03	No Suppression
	0		0.95	0.98			G	0.0E+00	Suppression
				0.02	0	0.0E+00	H	0.0E+00	Suppression
						1.0E+00	I	0.0E+00	No Suppression
					1	0.0E+00	J	0.0E+00	Suppression
						1.0E+00	K	0.0E+00	No Suppression
			0.05	0.98			L	0.0E+00	Suppression
				0.02	0	0.0E+00	M	0.0E+00	Suppression
						1.0E+00	N	0.0E+00	No Suppression
					1	0.0E+00	O	0.0E+00	Suppression
						1.0E+00	P	0.0E+00	No Suppression
						Non Suppression Probability		7.0E-03	

Fig. 11.9 Detection/suppression event tree solved at 15 min

When solved for 15 min, the resulting non-suppression probability is 0.007 (see Fig. 11.9), as at this point, the automatic sprinkler system has activated. The fire brigade has not arrived at this time. Finally, when solved for 25 min, the fire brigade has had 25 − 20 = 5 min available for fire control and suppression. This capability further lowers the non-suppression probability to 0.0042, as illustrated in Fig. 11.10.

Each of the three scenarios has a conditional probability of occurrence. Because each scenario represents a different end state, the conditional probability of each scenario should sum to 1.0. Table 11.4 summarizes these values. The cumulative suppression probability P_s is calculated as 1 minus the non-suppression probability P_{ns}. The conditional probability of Scenario 1, 0.65, is the probability that suppression occurs at 5 min or earlier, limiting the consequences to that level of consequence. During the subsequent interval, after 5 min but before 15 min (Scenario 2), the probability of suppression is the cumulative probability of suppression at 15 min minus the cumulative probability of suppression at 5 min, 0.993 − 0.65 = 0.343. A similar calculation applies to Scenario 3. This last interval receives the remaining probability, 1 − (0.65 + 0.342) = 0.0071.

In summary, there is a 1% chance of full consequences occurring given a hot work fire. The remaining 99% of the probability is apportioned among the first two scenarios. The resulting non-suppression probabilities are then incorporated in the scenario progression event tree to calculate the scenario's final risk values. For clarity purposes, the probabilities listed above in Table 11.4 can also be visualized in the form of a probability distribution where 65% represents the probability of suppression if T < 5. Similarly, 34% represents the probability of suppression if 5 < T < 15. Finally, 1% represents the probability of suppression if T > 15. Although there is a level of consequence defined at 25 min in this formulation, the value of 1% conservatively covers the remaining probability. Fig. 11.11 illustrates the probability

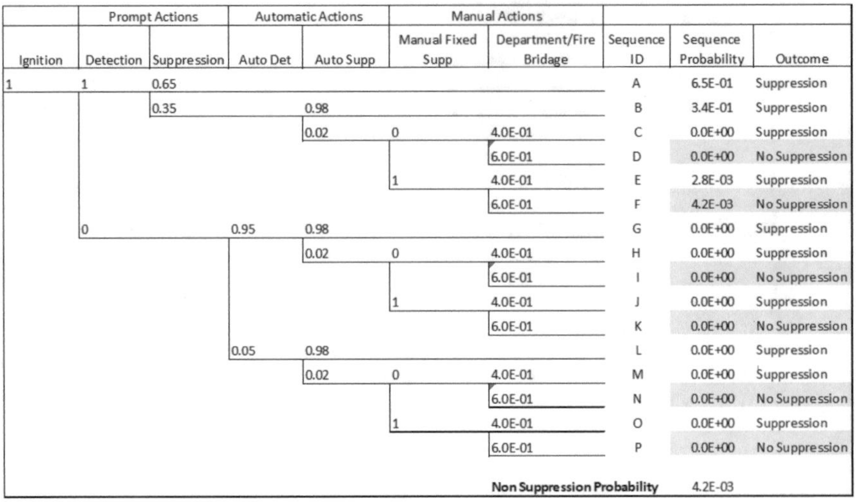

	Prompt Actions		Automatic Actions		Manual Actions				
Ignition	Detection	Suppression	Auto Det	Auto Supp	Manual Fixed Supp	Department/Fire Bridage	Sequence ID	Sequence Probability	Outcome
1	1	0.65					A	6.5E-01	Suppression
		0.35		0.98			B	3.4E-01	Suppression
				0.02	0	4.0E-01	C	0.0E+00	Suppression
						6.0E-01	D	0.0E+00	No Suppression
					1	4.0E-01	E	2.8E-03	Suppression
						6.0E-01	F	4.2E-03	No Suppression
	0		0.95	0.98			G	0.0E+00	Suppression
				0.02	0	4.0E-01	H	0.0E+00	Suppression
						6.0E-01	I	0.0E+00	No Suppression
					1	4.0E-01	J	0.0E+00	Suppression
						6.0E-01	K	0.0E+00	No Suppression
			0.05	0.98			L	0.0E+00	Suppression
				0.02	0	4.0E-01	M	0.0E+00	Suppression
						6.0E-01	N	0.0E+00	No Suppression
					1	4.0E-01	O	0.0E+00	Suppression
						6.0E-01	P	0.0E+00	No Suppression
						Non Suppression Probability		4.2E-03	

Fig. 11.10 Detection/suppression event tree solved at 25 min

Table 11.4 Summary of conditional probabilities for each event in the chronology

Scenario	Elapsed time after fire ignition (min)	Estimated non-suppression probability, P_{ns}	Cumulative suppression probability, $P_s = 1 - P_{ns}$	Formula for the conditional probability of the scenario	Conditional probability of the scenario
1	5	0.35	0.65	$P_s(5)$	0.65
2	15	0.007	0.993	$P_s(15) - P_s(5)$	0.343
3	>15	0.0042		*Remaining probability*	0.007

distribution (i.e., which includes scenarios 4 and 5 in the scenario progression event tree).

Figure 11.12 illustrates the scenario progression event tree, including the conditional probabilities for detection and suppression. The event tree also includes the resulting scenario frequencies resulting from the ignition. Notice that the normalization process suggests a probability of 0.01 for the event postulated at 25 min (i.e., brigade arrival). This is reflected in the event tree as the combination of both the success and failure of the fire brigade.

Notice that in Fig. 11.12, a value of 1.0 is assigned to the tree's lower branches due to the normalized suppression probabilities assigned. For clarity purposes, the event tree can be configured as illustrated in Fig. 11.13, which depicts each outcome of the tree with the respective probabilities, which add to 1.0.

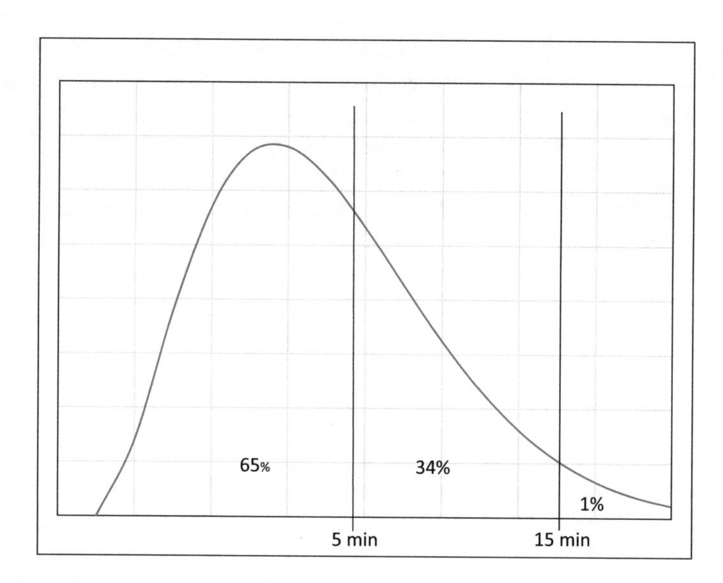

Fig. 11.11 Distribution illustrating the probability of suppression for each point in time (Figure not to scale)

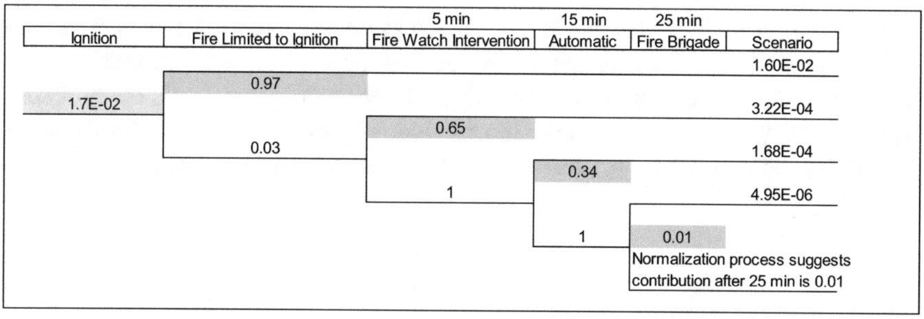

Ignition	Fire Limited to Ignition	5 min Fire Watch Intervention	15 min Automatic	25 min Fire Brigade	Scenario
1.7E-02	0.97				1.60E-02
	0.03	0.65			3.22E-04
		1	0.34		1.68E-04
			1	0.01 Normalization process suggests contribution after 25 min is 0.01	4.95E-06

Fig. 11.12 Event tree for the example with detection and suppression probabilities for each event in the chronology

Ignition	Fire Limited to Ignition Source	Detection and Suppression		Scenario Frequencies
1.7E-02	0.97			1.60E-02
	0.03	0.65	5 min: Fire watch intervention	3.22E-04
		0.342	15 min: Automatic Sprinklers	1.69E-04
		0.01	25 min: Fire Brigade Intervention	4.95E-06

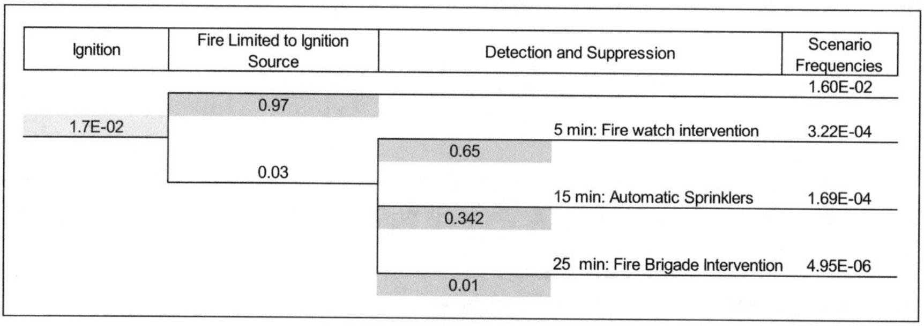

Fig. 11.13 Event tree for the example with detection and suppression probabilities for each event in the chronology

The calculated suppression probabilities can also be calculated using conditional probabilities, which would produce identical results. Each branch requires the determination of the probability of non-suppression that is dependent on a previous suppression credit. In the event tree, suppression is credited at multiple points along the scenario progression. Therefore, the resulting probabilities are dependent on any previously credited suppression. This is calculated using a conditional probability as shown in Eq. 11.5.

$$P(B \mid A) = \frac{P(A \cap B)}{P(A)} \tag{11.5}$$

where:

Event A = the previous suppression credit
Event B = the current event suppression credit
$P(B|A)$ = the conditional probability of Event B occurring given that Event A has already occurred
$P(A \cap B)$ i = the intersection of Event A and Event B, which is the probability that both events occur
$P(A)$ = the probability that Event A has occurred.

In this context, Events A and B are defined as non-suppression events. The term $P(A \cap B) = P_{ns}(t_B)$ is the total probability of non-suppression at the time of the second event (i.e., the current event in time, t_B). The term $P(A) = P_{ns}(t_A)$ is the probability of non-suppression at the time of the previous event (i.e., the time of Event A, t_A). Therefore, the dependent suppression function is shown in Eq. 11.6. [5]

$$P(B \mid A) = \frac{P_{ns}(t_B)}{P_{ns}(t_A)} \tag{11.6}$$

This allows for multiple suppression events to be credited along any single branch path of the event tree. In all cases, the total suppression credit ensures that the total credit for suppression is equivalent through the event tree. For example, the non-suppression probability for the event at 15 min is:

$$P(B \mid A) = \frac{P_{ns}(15\,\text{min})}{P_{ns}(5\,\text{min})} = \frac{0.007}{0.342} = 0.02$$

Using this value, the third scenario sequence can be calculated as $1.7E-2 \times 0.03 \times 0.342 \times (1-0.02) = 1.68E-4$, which is equivalent to the evaluation presented in Fig. 11.12.

Modeling Egress in a Fire Risk Assessment

The conceptual example associated with hot work hazards discussed so far has focused on detection and suppression capabilities represented chronologically in the event tree. Egress capabilities can be implicitly evaluated as part of the chronology of the event tree by expressing the occupant's ability to evacuate the facility in the consequence term. Each point in time captured in the chronology of the tree is associated with a fire size that will challenge the occupant's ability to evacuate. For each postulated fire scenario, egress capabilities will need to be evaluated for effectiveness and availability (e.g., as suggested using the fire scenario concept tree, code compliance evaluations, or performance-based egress modeling) to assess the consequence level at each point in the chronology.

11.2.2 Consequence Assessment

Chapter 10 included background information on the factors influencing the assignment of consequences. That information is applicable for both qualitative and quantitative assessments. Applying those concepts in the hot works example discussed in this chapter, the tree structure may suggest increasing levels of consequences as suppression attempts fail. Each of these consequence levels should consider life safety (injuries or fatalities), property protection, environmental impact, and business continuity, depending on the source of ignition, which may be generated even for non-propagating fires. Specifically, the following consequence levels are assigned, as depicted in Fig. 10.9.

- *Consequence level 1:* A hot work fire that is controlled by fire prevention measures and does not propagate to nearby combustibles will have negligible consequences. Numerically, Table 7.3 in Chap. 7 recommends a value of 1.0E-5 as a normalized consequence term.
- *Consequence level 2:* The fire watch successfully detects and promptly controls the fire using a fire extinguisher. Such a rapid response is expected to have negligible to marginal consequences. Numerically, Table 7.3 in Chap. 7 recommends a value of 1.0E-4 as a normalized consequence term.
- *Consequence level 3:* A fire that grows to an intensity where the fire watch is no longer capable of controlling it and, at the same time, capable of activating a sprinkler system may produce consequences larger than marginal. This will depend on the nature of the facility, the losses associated with the ignition source, and nearby combustible and life safety considerations. A value of 1.0E-2 is assigned based on the values in Chap. 7 Table 7.3.
- *Consequence Level 4 and Level 5:* These are associated with sprinkler failures and rely on the fire brigade or fire department to control the fire before flashover conditions. A fire resulting in flashover conditions is often associated with consequences ranging from major to catastrophic. A value of 1.0 is assigned. This value captures the midpoint in the range of major to catastrophic.

With these consequence assignments, the fire risk evaluation can now proceed. Figure 11.14 depicts the consequences and the risk quantification for each sequence in the event tree.

Ignition	Fire Limited to Ignition Source	Detection and Suppression	Scenario Frequencies	Consequences		Risk
1.7E-02	0.97		1.60E-02	1.00E-05	1- Negligible	1.6E-07
	0.03	0.65 5 min: Fire watch intervention	3.22E-04	1.00E-04	2- Negligible to Marginal	3.2E-08
		0.342 15 min: Automatic Sprinklers	1.69E-04	1.00E-02	3- Marginal or Major	1.7E-06
		0.01 25 min: Fire Brigade Interventior	4.95E-06	1.00E+00	4- Major to Catastrophic	5.0E-06

Fig. 11.14 Scenario progression event tree for hot work example with assessment for consequences

11.3 Risk Estimation

Once consequence levels are assigned to each of the scenario's potential outcomes, the risk matrix is used to determine the level of risk associated with each scenario. The process consists of identifying the risk level specified in the matrix corresponding to the scenario frequency and consequence level assigned.

The hot work example that has been discussed in this chapter has five possible outcomes. Each of those is treated as an individual scenario. Each has an ignition frequency, a scenario frequency characterized by fire protection capabilities numerically evaluated with a conditional probability and corresponding consequences. To do so, the risk evaluation process assesses each of these five scenarios using the qualitative risk matrix identified in Table 7.4 (Table 11.5).

Table 11.6 summarizes the risk evaluation results. In this example, the outcome of each sequence is associated with an acceptable level of risk. The risk provides the following insights:

- *The importance of fire prevention practices.* The conditional probability for fire severity captures the barrier in place between hot work and ordinary combustibles. Such practices influence the risk numbers and should be monitored (see Chap. 15) throughout the facility's operational life.
- *Importance of automatic sprinklers.* This system has the practical effect of lowering the scenario frequency, which suggests that the risk is acceptable depending on the specific characteristics of the scenario. For example,

  (a) If consequences are low, the resulting risk may be tolerable or acceptable based on the loss associated with the postulated scenario.

  (b) In the case of high consequences based on the loss associated with the postulated scenario, further evaluation may be necessary to ensure enough margin in the results so that in-depth defense strategies are available.

- If an automatic sprinkler system is not available or deemed ineffective, the scenario frequency may not be as low. This suggests the need to improve and moni-

Table 11.5 Quantitative risk matrix

		Consequence				
		Negligible	Marginal	Major	Critical	Catastrophic
Frequency		1.0E-06	1.0E-04	1.0E-02	5.0E-01	1.0E+00
Frequent	1.0E+00	1.0E-06	1.0E-04	1.0E-02	5.0E-01	1.0E+00
Probable	1.0E-01	1.0E-07	1.0E-05	1.0E-03	5.0E-02	1.0E-01
Occasional	1.0E-02	1.0E-08`	1.0E-06	1.0E-04	5.0E-03	1.0E-02
Remote	1.0E-04	1.0E-10	1.0E-08	1.0E-06	5.0E-05	1.0E-04
Improbable	1.0E-06	1.0E-12	1.0E-10	1.0E-08	5.0E-07	1.0E-06
Incredible	1.0E-08	1.0E-14	1.0E-12	1.0E-10	5.0E-09	1.0E-08

Table 11.6 Summary of risk estimation results for hot work example

Scenario	Ignition frequency (initiating event)	Scenario frequency	Consequences	Risk evaluation	Comment
1	1.7E-2	1.6E-2	1.0E-5, Negligible	1.6E-7, Acceptable	Fire prevention measures prevent propagation
2	1.7E-2	3.22E-4	1.0E-4, Marginal	3.22E-8, Acceptable	Fire prevention and hot work fire watch credited in the analysis. Distance from hot work activities to ordinary combustibles is also credited in the analysis.
3	1.7E-2	1.69E-4	1.0E-2 Marginal to major	1.7E-6, Acceptable	Automatic sprinklers are credited in the analysis. Distance from hot work activities to ordinary combustibles is also credited in the analysis.
4 & 5	1.7E-2	4.95E-6	1.0 Major to catastrophic	4.0E-6, Acceptable	Fire brigade or fire department credited for suppression. Distance from hot work activities to ordinary combustibles is also credited in the analysis.

tor fire prevention strategies, the fire watch capability, and the response time of the fire brigade if these capabilities can control the identified fire hazards.

- The concept of availability characterizing the fire protection capabilities in the analysis should be used to evaluate the level of scenario frequency reduction assigned to the scenario. For example, a scenario frequency reduction from occasional to remote may not be assigned if the sprinkler system is routinely out of service.

Although this chapter described the process for systematic quantitative risk estimation for a single ignition associated with hot work hazard developing into five fire scenarios, a full fire risk assessment should include similar evaluations for all the identified scenarios within the scope of the analysis and a risk evaluation for the cumulative risk in the facility. Each scenario will be characterized by its risk contribution to the overall risk of the facility. Chapter 12 provides further details about risk quantification and evaluation.

11.4 Dealing with the Normalized Consequence Term

The hot work example presented earlier in this chapter can be used to describe how to deal with the normalized consequence term. For this description, it is assumed that a maximum monetary loss of $10,000,000 for a major or catastrophic consequence can be realized. Table 11.7 lists the scenario number and the conditional probability (i.e., scenario frequency/ignition frequency), normalized consequences, and the consequences in monetary terms ($/Fire). The risk, expressed in $/Fire, is the multiplication of the ignition frequency, the scenario frequency, and the consequence term.

Adding the risk of each branch in the scenario progression event tree results in the total annualized loss per hot work ignition. Furthermore, adding the annualized value for all the scenarios included in a comprehensive risk assessment will result in the total annualized expected loss for the facility. Therefore, a total annualized loss per individual hot work scenario progression is estimated to be $68.4. This value can be further calculated for a predetermined period of time (insurance period). For a period of 10 years, the maximum probable loss (i.e., the largest loss that an insurance policyholder could expect to experience if a certain event occurred) would be $684.00, assuming no interest effects over time for this fire scenario. This expected probable loss is based on the features of the fire protection program incorporated in the risk assessment. It is noted that removing conditional probabilities representing the fire protection features would yield a conservative value that may represent the maximum foreseeable loss (the highest loss that can possibly happen to an insured).

To illustrate the difference between the maximum expected loss and the maximum foreseeable loss, the example described above in Table 11.7 is reevaluated without the effects of the sprinkler system. Removing the sprinkler system

Table 11.7 Example of estimated annualized loss per fire scenario

Scenario	Ignition frequency (fires/year)	Conditional probability	Normalized consequence	Loss ($/Fire)	Risk ($/year)
1	1.7E-02	9.70E-01	1.00E-05	$100	$1.60
2	1.7E-02	1.95E-02	1.00E-04	$1000	$0.32
3	1.7E-02	1.03E-02	1.00E-02	$100,000	$16.93
4	1.7E-02	3.00E-04	1.00E+00	$10,000,000	$49.50
Total annualized loss per hot work scenario:					$68.4

Table 11.8 Summary of conditional probabilities for each event in the chronology assuming no automatic sprinkler system

Scenario	Elapsed time after fire ignition (min)	Estimated non-suppression probability, P_{ns}	Cumulative suppression probability, $P_s = 1 - P_{ns}$	Formula for the conditional probability of the scenario	Conditional probability of the scenario
1	5	0.35	0.65	$P_s(5)$	0.65
2	15	0.35	0.65	$P_s(15) - P_s(5)$	0.000
3	>15	0.21	0.79	*Remaining probability*	0.35

Table 11.9 Example of estimated annualized loss per fire scenario

Ignition frequency	Conditional probability	Normalized consequence	Loss ($)	Risk ($/year)
1.7E-02	9.70E-01	1.00E-05	$100	$1.65
1.7E-02	6.50E-01	1.00E-04	$1000	$11.05
1.7E-02	0.00E+00	1.00E-02	$100,000	$0.00
1.7E-02	3.50E-01	1.00E+00	$10,000,000	$59,500.00
			Total annualized loss	**$59,512.7**

suppression capability and updating the analysis with identical values (except setting the capability for automatic sprinklers to FALSE in Table 11.3), the normalized probability of catastrophic consequences significantly increases to 0.35. The normalization is summarized in Table 11.8. It is noted that this normalization process has been described earlier in this chapter in Table 11.4.

With the updated non-suppression probabilities calculated without including the effects of the automatic sprinkler system in the scenario, the total annualized loss in this example fire scenario is, as expected, significantly larger with a value of $59,512 per year (Table 11.9).

References

1. U.S. Nuclear Regulatory Commission, *Hot Work Transient Fire Frequency Influence Factors. FAQ-12-0064, R1*, [Online]. Available: https://www.nrc.gov/docs/
2. U.S. Nuclear Regulatory Commission, *NUREG 2232: Heat Release Rate and Fire Characteristics of Fuels Representative of Typical Transient Fire Events in Nuclear Power Plants* (U.S. Nuclear Regulatory Commission, Washington, DC, 2020)
3. U.S. Nuclear Regulatory Commission, *NUREG-2233: Methodology for Modeling Transient Fires in Nuclear Power Plant Fire Probabilistic Risk Assessment* (U.S. Nuclear Regulatory Commission, Washington, DC, 2020)
4. U.S. Nuclear Regulatory Commission/EPRI, *Refining and Characterizing Heat Release Rates from Electrical Enclosures During Fire, Volume 1: Peak Heat Release Rates and Effect*

of Obstructed Plume, NUREG-2178, vol 1 (U.S. Nuclear Regulatory Commission/EPRI, Washington, DC/Palo Alto, 2016)
5. U.S. Regulatory Commission/EPRI, *Refining and Characterizing Heat Release Rates from Electrical Enclosures During Fire, Volume 2: Fire Modeling Guidance for Electrical Cabinets, Electric Motors, Indoor Dry Transformers, and the Main Control Board, NUREG-2178*, vol 2 (U.S. Regulatory Commission/EPRI, Washington, DC/Palo Alto, 2019)
6. U.S. Nuclear Regulatory Commission/EPRI, *Nuclear Power Plant Fire Ignition Frequency and Non-suppression Probability Estimation Using the Updated Fire Events Database: United States Fire Event Experience Through 2009, NUREG-2169* (U.S. Nuclear Regulatory Commission/EPRI, Washington, DC/Palo Alto, 2014)
7. U.S. Nuclear Regulatory Commission/EPRI, *EPRI/NRC-RES, Fire Human Reliability Analysis Guidelines. NUREG-1921* (U.S. Nuclear Regulatory Commission/EPRI, Washington, DC/Palo Alto, 2012)

Chapter 12
Risk Evaluation

Risk evaluation and acceptability refers to comparing the assessed risk for an individual or a group of scenarios with an acceptance criterion typically defined in the risk matrix and agreed upon by stakeholders. Risk could also be evaluated against a reference value of individual and societal risk derived for a code-compliant building, facility, or process. In this case, the resulting risk should not exceed the reference value, which is implicitly considered acceptable based on code compliance or using the ALARP principle described earlier in Chap. 7. The comparison is represented as a decision point in the process of determining if the risk associated with a given scenario(s) is "acceptable" or "tolerable." Examples of the risk evaluation process were provided earlier in Tables 10.1 and 11.6. These tables include an assessment of the acceptability of risk results using a risk matrix as the tolerability or acceptance criteria.

The risk evaluation process can lead to the following:

- The assessed risk is low enough that further analysis and refinements are not warranted.
- The assessed risk is understood to be a conservative estimate suggesting that further analysis, such as incorporating available fire protection alternatives or removing conservatisms, should be considered before a final decision.
- The assessed risk reflects a detailed analysis of most conservatisms refined to the extent practical. In this case, risk values exceeding the acceptability thresholds may indicate that modifications or improvements in the fire protection strategy are necessary to improve safety.

In practice, the above outcomes are the basis for the iterative nature of the risk evaluation process. Typically, the initial stages often consist of using easily obtained inputs and may ignore risk-influencing factors that may reduce risk to simplify obtaining preliminary results. This early stage suggests which scenarios have low risk with a relatively small analytical effort, reducing the number of scenarios where further analysis is necessary.

SFPE Guide to Fire Risk Assessment, The Society of Fire Protection Engineers Series, https://doi.org/10.1007/978-3-031-17700-2_12

The objectives of the analysis govern the evaluation process. For example,

- A risk assessment developed to evaluate and select a fire protection design option may require a detailed representation of different alternatives such that the risk estimates can be adequately compared. Conservatively ignoring fire protection design options may identify scenarios in which a given design is unnecessary but will not have the resolution necessary to compare different options.
- An analysis developed for evaluating the risk associated with fire protection deficiencies in operating facilities (e.g., lack of code compliance in a specific configuration) may take advantage of conservative modeling to characterize the level of safety with the assessed risk. For example, not considering the effects of detection and suppression in a scenario may provide conservative risk results without incorporating these elements in the analysis. If the risk is acceptable, further analysis may not be necessary (e.g., a scenario with low frequency and/or low consequences). If the risk is not acceptable, further analysis may be necessary for reflecting detection or suppression capabilities in the fire risk estimates.
- An AHJ may request conservative modeling of the scenario to ensure a safety margin in the analysis. Such analysis would be developed to address regulatory interactions and may not support other risk-informed applications.
- An analysis developed to support financial and operational decisions may require the removal of conservatism and rely on realistic input parameters and assumptions.
- An analysis focused on determining annualized loss for insurance purposes. In this case, realistic risk results may need to reflect monetary losses in the consequence term.

Again, the total risk is the sum of the risk of all the individual scenarios within the scope of the analysis. In some applications, the total number of scenarios may be relatively large. Therefore, an iterative process intended to identify the detail required for modeling each scenario becomes a valuable and practical tool in the process. At the same time, such a process should be implemented to minimize the potential for overestimating or underestimating risk. Conceptually, consider a relatively large number of individual scenarios that have been conservatively evaluated, resulting in low risk. Although individually they are low risk, cumulatively, the scenarios may represent a risk level that may not be low. Consider the "hot work" scenarios resulting from one postulated ignition evaluated in the previous chapter that resulted in acceptable risk. Cumulatively, the sum of the risk associated with all hot work scenarios throughout an entire facility may suggest a risk level requiring improvements in the fire protection strategy (e.g., improving the hot work procedures or prohibiting hot work activities in specific locations) even if individually each scenario has an acceptable risk.

Although conservative risk modeling is a necessary strategy for developing a risk assessment cost-effectively, it may "mask" the effects of fire protection strategies and limit the insights obtained from the analysis. Recall that risk values are relative and may only provide insights promoting fire safety when compared to each other. Therefore, conservative modeling may not reflect the risk reductions associated with specific fire protection features or strategies, which may reduce the effectiveness of the analysis in supporting day-to-day fire protection decision-making. Ideally, the iterative risk assessment process ends when:

1. The analysis reflects in detail the fire protection capabilities and procedures for the top risk contributing scenarios so that any remaining scenario that is conservatively modeled has a risk lower than those that are analyzed in detail, and.
2. The total risk (i.e., the sum of the risk associated with individual scenarios) is acceptable or tolerable. Such a strategy minimizes the potential of masking important fire safety insights.

As suggested earlier, the risk evaluation process supports the decision-making process associated with risk acceptability. Acceptability decisions can be made in terms of criteria agreed upon by stakeholders, the cost associated with any evaluated option/process/design that delivers the greatest risk reduction, consideration of the robustness of any evaluated design option/process for reliance on any single risk mitigation measure, or a comparison with a base case configuration representing an acceptable risk level.

Chapter 13
Sensitivity and Uncertainty Analysis

Applying a fire risk assessment requires simplifying assumptions, the use of limited data, elicitation of engineering judgment, and the use of analytical or empirical models. In addition, sometimes conservative assumptions and input variables are used to ensure a margin of safety in the results (e.g., an AHJ may prefer or request an analysis based on reasonable worst-case conservatism rather than the assessment of realistic risk results). These elements unavoidably introduce uncertainties into the analysis. Therefore, a comprehensive step focusing on sensitivity and uncertainty analysis is necessary to assess the impact of these analytical decisions on the study results so that conclusions and recommendations are made considering their effects.

A sensitivity and uncertainty analysis may provide valuable input into how a specific design or procedure could influence fire safety for new designs. For existing designs, a sensitivity and uncertainty analysis can help determine which systems need to be monitored to ensure the risk assessment results are maintained to preserve the estimated levels of risk. It can also assist in risk informing procedure changes and modifications associated with the facility operation.

Although this step is often performed in the final stages of the analysis, identifying and characterization of sources of uncertainty should be part of the risk assessment development process. Similarly, determining any sensitivity to the inputs and assumptions made in the study could have implications for the duration of the life of the subject building, facility, or process.

The inputs to this task are all the sources of uncertainties inherent in the previous tasks. These sources may come from the following:

- Governing assumptions made throughout the analysis
- Characterization of the inputs (data-driven, based on judgment, or analytical decision based on environmental conditions or material properties) to the risk assessment
- Models (analytical, empirical, statistical, etc.) used for assessing consequences

The output of this task is an assessment of the uncertainty or sensitivity of the above elements concerning the final results. This is expressed as a metric on how

SFPE Guide to Fire Risk Assessment, The Society of Fire Protection Engineers Series, https://doi.org/10.1007/978-3-031-17700-2_13

much the individual elements affect the final results and subsequent decisions. The uncertainty and sensitivity assessment results may also be used as input supporting a risk monitoring program (See Chap. 15).

13.1 Sensitivity vs. Uncertainty

A fire risk assessment may not be sensitive to an uncertain variable. Similarly, a sensitive parameter in a model may not be uncertain. The uncertainty in a variable represents the lack of knowledge associated with the variable. Uncertainty is often expressed in probabilistic risk assessment with probability distributions. In contrast, a variable's sensitivity in a model is defined as the rate of change in the model output with respect to changes in the variable.

13.1.1 Sensitivity Analysis

Sensitivity analysis refers to evaluating model outputs given variations in the different input parameters to the model. In other words, sensitivity refers to the rate of change of the model output with respect to input variations. A sensitivity analysis can determine if changes in an element result in a difference in the output (risk estimation) significantly enough to change the outcome of decisions based on the assessment. In general terms, a sensitivity analysis can provide a robust and relatively simple approach for dealing with uncertainties. The sensitivity analysis does not suggest the uncertainty in the risk values. Given the uncertainties in the inputs and the modeling approach, it will identify sensitive variables for which the uncertainty should be minimized.

13.1.2 Uncertainty Analysis

Uncertainty analysis refers to the identification and propagation of uncertainties throughout the fire risk assessment. The objective of the uncertainty analysis is to assess the variability in the model output given uncertainties in the input values, governing assumptions, and implemented models.

Theoretically, uncertainty is classified as epistemic and aleatory. Epistemic uncertainty or knowledge uncertainty is the type that can be reduced with additional research and/or resources. On the other hand, aleatory uncertainty, or variability, reflects the inherent randomness of the parameter and cannot be reduced. In practice, these uncertainties are often treated in combination in the form of parameter uncertainty, model uncertainty, and completeness uncertainty. Parameter uncertainty refers to the uncertainty associated with the input parameters in the analysis.

On the other hand, model uncertainty refers to the uncertainties related to the structure or mathematical formulation of the models used within the risk assessment. Finally, completeness uncertainty is associated with phenomena that may not be fully captured in the analysis.

13.2 Sensitivity and Uncertainty in Practice

The flowchart in Fig. 13.1 presents a general overview of a sensitivity and uncertainty analysis. This flowchart briefly introduces each activity identified in each subsection.

In general, the process consists of identifying the elements in the assessment that are uncertain for which the results are sensitive. The term element refers here to assumptions, input parameters, or models used in the analysis. This process includes:

- Listing assumptions, input parameters, and models used in the risk assessment and qualitatively evaluate the need for sensitivity or level of uncertainty associated with those elements.

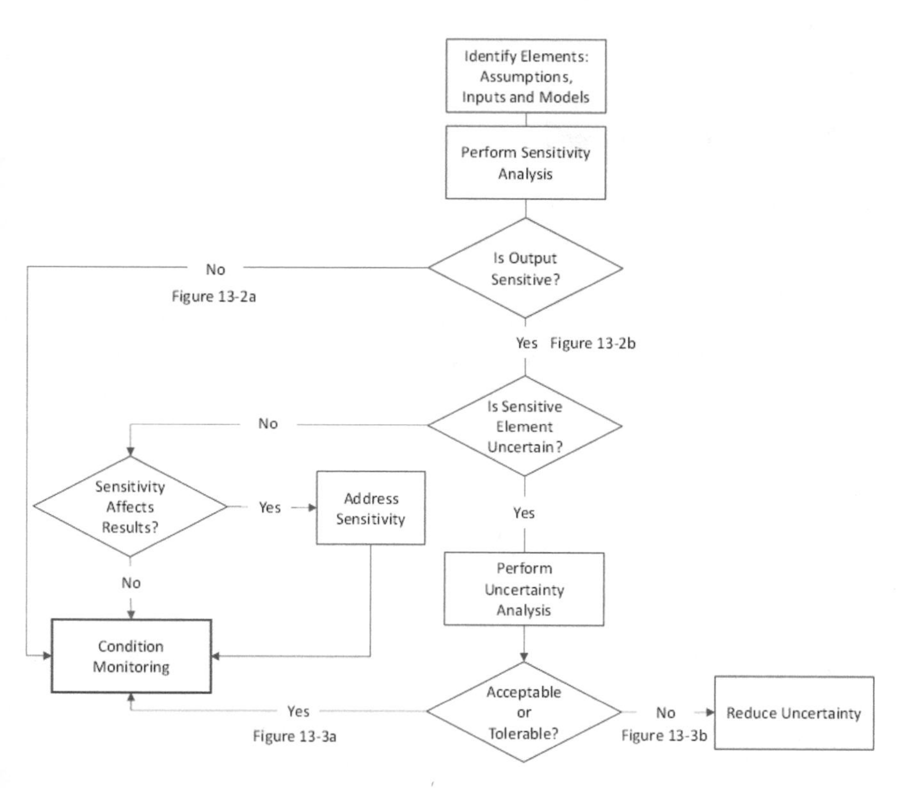

Fig. 13.1 Sensitivity and uncertainty analysis flow chart

- Once the qualitative assessment is performed, the elements are evaluated in a sensitivity analysis to determine their impact on the results. If the results are not sensitive to variations, the element is referred to the monitoring program. On the other hand, a sensitivity range that impacts the results and conclusion of the analysis should be evaluated to determine their level of uncertainty.
- If the uncertainty is low (e.g., well-known inputs), the analyst may need to evaluate design improvements to monitor the design conditions. If the uncertainty analysis suggests that the results and conclusions will not be affected by the uncertainty associated with the assumptions, input, or model, no further analysis is necessary, and these governing elements are then referred to the monitoring program. Alternatively, if the results or conclusions of the assessment are impacted by the uncertainty associated with the given element, recommendations may be necessary for further research and analysis to reduce the uncertainty in the given element.

Additional details on the process are summarized later in this section. It is noted that the monitoring program is the final step of the sensitivity and uncertainty analysis. Assumptions, input parameters, and models governing the risk assessment results need to be monitored and maintained throughout the facility's operational life to ensure that the results are valid. See Chap. 15 for additional information on risk monitoring.

13.2.1 Identify Elements

As noted above, the inputs to this task are the elements included and identified in all the previous tasks. At a high level, these elements include:

- Governing assumptions (e.g., 3-h rated firewalls will prevent fire propagation from one side of the boundary to the other).
- Input values to the risk assessment (e.g., ignition frequencies, ambient conditions, damage criteria, fuel source properties, fire sizes, fire durations, characterization of detection and suppression features).
- Models (analytical, empirical, statistical, etc.) used for assessing consequences.

The listed set of assumptions, input values, and models define the sensitivity and uncertainty analysis scope to be performed.

13.2.2 Sensitivity Analysis

Once all the elements have been identified, a determination is made on specific sensitivity cases to understand the impact on the final results.

13.2.2.1 Sensitivity Analysis Process

Sensitivity analysis involves solving the risk model using a range of values assigned to the elements of interest to represent potential outcomes. This is not limited to varying input parameters in the conservative direction, as often input parameters in the base model have been already conservatively assigned. In such cases, realistic values in the sensitivity analysis may provide valuable insights supporting the decision-making process. The analysis may be performed in one of two ways:

- An element may be varied by an appropriate range or distribution of values, or.
- An element may be removed (or added).

In the first method, one variable at a time is altered with a fixed fraction of the variable's basic value, for example, by 50%. Usually, the value is increased and decreased to determine the impact of the change. A large change in the output value with respect to using the basic value (base risk), compared to the effect of changing other variables, indicates a sensitive variable. In the second method, the sensitivity of an element on the results is determined by simply adding or removing the element from the analysis. For example, the sensitivity of a wet-pipe sprinkler system may be estimated by first including the system and then removing the system from the analysis (e.g., setting the failure probability to 1.0 in a quantitative assessment). Examples of elements that may lend themselves to a credit/no credit sensitivity analysis include:

- Crediting active fire protection systems (detection, suppression, manual/fire brigade suppression actions, etc.)
- Crediting passive fire protection systems (firewalls, fire doors, etc.)
- Crediting procedures (limiting combustible materials in specific locations, etc.)

Examples of elements for which the sensitivity may be assessed using a range or distribution of values:

- Ignition frequency
- Fire size
- Ambient conditions (temperature, humidity, etc.)
- Material properties (thermal conductivity, specific heat, density, etc.)
- Fuel properties (mass burning rate, heat of combustion, density, soot yield, etc.)
- Equipment reliability and unavailability (detection, suppression systems)
- Human error probabilities
- Timing (detection, suppression, brigade arrival times)
- Effects of procedures (may influence the range or distribution of elements listed above)

13.2.2.2 Sensitivity Analysis Results

Once performed, the sensitivity analysis suggests changes in an element result in a change in the output (risk estimation), which is significant enough to change decisions based on the assessment. Figure 13.2 illustrates sensitive and insensitive outputs to changes in an element used in the fire risk assessment. In this figure, the term "base risk" refers to the results of the base case assessment where input values have not been replaced with those in the sensitivity case. The term "sensitivity risk" refers to results obtained after varying replacing based values with those selected for the sensitivity analysis.

If results are found to be insensitive to changes in the elements (See Fig. 13.2a), then, in most cases, no additional analysis is required. Future monitoring is required of the elements included in the assessment that do not change to ensure the results remain applicable going forward. See Chap. 15 for additional information.

If the results are sensitive to changes in the element, but such sensitivity does not affect outcome or decisions, no additional analysis would be required, but conditions should be monitored to ensure the results remain applicable in the future.

Alternatively, if a change in the risk associated within an element results in a change in the outcome/decision reached by the assessment (See Fig. 13.2b), the next step is to determine the level of uncertainty associated with such an element.

13.2.3 Uncertainty Analysis

As suggested earlier, once it has been determined that the risk assessment results are sensitive to an element, the analyst should identify the uncertainty associated with that element. If the element is not uncertain, the risk assessment should then provide relevant insights and information. On the other hand, if the element is uncertain, further evaluation may be necessary to identify if the uncertainty can affect the

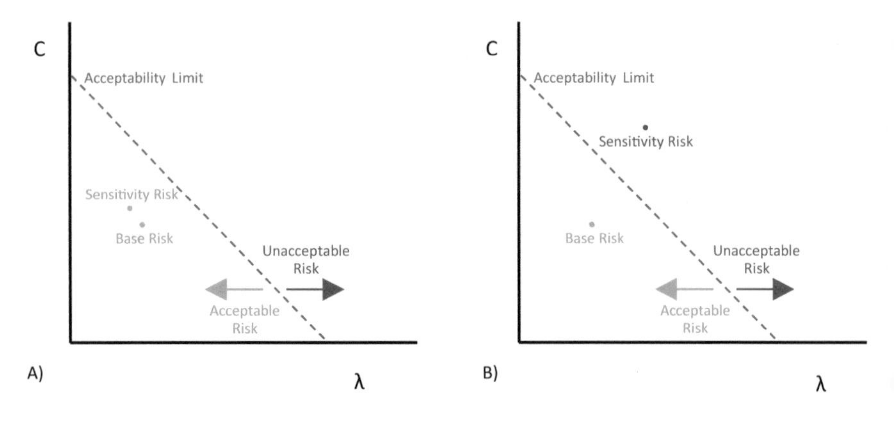

Fig. 13.2 Results shown to be (**a**) insensitive and (**b**) sensitive to changes in elements

decision-making process. An uncertainty analysis involves propagating uncertain elements through the model and observing the effect on the results.

13.2.3.1 Uncertainty Analysis Process

Three types of uncertainty should be considered as summarized below:

Parameter uncertainty [1]: Analysis of uncertainty in parameters involves choosing values from statistical distributions or estimating generic reference data and propagating these values through the assessment to observe the resulting uncertainty in the model prediction. The process of assessing the uncertainty in a parameter is similar to that of a sensitivity analysis.

Model uncertainty [1]**:** Models attempt to represent reality. By doing so, they often rely on idealizations of physical phenomena and simplifying approximations. Uncertainty (and bias) is estimated through the process of verification and validation (V&V). Verification is the process of determining that the calculation method accurately represents the developer's conceptual description of the calculation/solution, and validation is the process of determining the degree to which a calculation method is an accurate representation of the real world from the perspective of the intended uses of the calculation method. Part of the validation involves recognizing what biases and uncertainties are included in the model's representation of reality. Examples of fire simulation models that specifically document verification and validation (V&V), as well as discussions of model bias and uncertainty, including the Consolidated Model of Fire and Smoke Transport (CFAST) and Fire Dynamic Simulator (FDS) software [2]. The bias and uncertainty associated with popular hand calculations are also discussed in Supplement 1 to NUREG-1824 [3]. While these discussions are developed explicitly for fire modeling in nuclear power plants, the modeling application could be applied across any industry.

Completeness uncertainty [1]: Completeness refers to the likelihood that a model may not completely describe the phenomena it was developed to predict, given any inherent simplifications and approximations. This uncertainty is often accounted for in the estimations of model uncertainty. Alternatively, completeness could refer to the possibility that certain fire hazards should not be identified or specific failure modes are not considered during a fire risk assessment.

Once ranges or distributions for elements of interest are identified or developed, these values are propagated through the model, and the effect on the risk results are characterized.

13.2.3.2 Uncertainty Analysis Results

In Fig. 13.3, examples of outputs that have been found to have uncertain results that do not result in a change in the decisions concluded in a fire risk assessment (a) and results that likely would result in a need to change the risk assessment results (b) are presented. The uncertainty is represented by the variability in the results depicted by

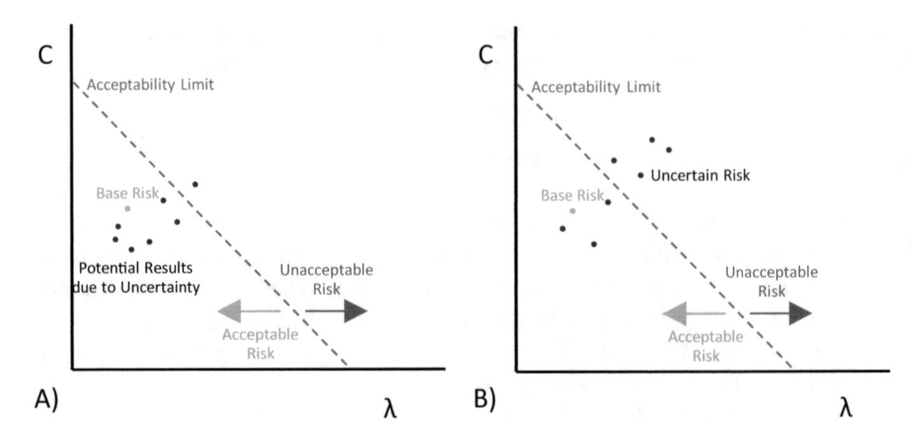

Fig. 13.3 (**a**) Acceptable and (**b**) unacceptable uncertain results

the various potential solutions. Increased scatter in the solutions is an indication of a higher degree of uncertainty.

Figure 13.3a presents an example of uncertain results that suggest that the decision-making process should not be impacted. Despite being uncertain, the outcome remains within the same range of tolerance or acceptability.

The range of results presented in Figure 13.3b is an example of an uncertainty range that may impact the decision-making process. No decision can be made given these results, as the possible outcomes are spread widely across acceptable/tolerable limits. When this is the outcome of the uncertainty analysis, the next step is to identify and revise the elements that are driving the results until the uncertainty is reduced.

References

1. U.S. Nuclear Regulatory Commission, and Electric Power Research Institute (EPRI), *Nuclear Power Plant Fire Modeling Analysis Guidelines, NUREG-1934 and EPRI 1023259* (U.S. Nuclear Regulatory Commission, and Electric Power Research Institute (EPRI), Washington, DC and Palo Alto, 2012)
2. ISO, *16730-1: Fire Safety Engineering—Procedures and Requirements for Verification and Validation of Calculation Methods—Part 1: General* (ISO, Geneva, 2015)
3. M. Salley, *15. NUREG-1824, Supplement 1, "Verification and Validation of Selected Fire Models for Nuclear Power Plant Applications: Draft Report," NUREG-1824/EPRI 3002002182* (NRC, Washington, DC, 2014)

Chapter 14
Documentation and Risk Communication

Once the evaluation process is completed, the risk assessment should be adequately documented and communicated to the stakeholders so it can be maintained and monitored throughout the facility's life cycle. The following elements of the analysis should be documented and communicated appropriately to the corresponding stakeholders:

- Objectives and scope of the fire risk assessment
- Assumptions and limitations influencing the evaluation of the assessment
- The acceptability or tolerability criteria governing the decision-making process
- The fire risk assessment methodology
- The identified hazards and fire scenarios
- The risk estimation
- Risk evaluation and acceptability
- The analysis results, including a characterization of the frequency and consequences of the fire scenarios within the scope of the study
- The results from the uncertainty and sensitivity analysis
- Quality assurance (internal and external review)
- Assessment of applicable fire risk assessment standard requirements
- The conclusions of the assessment, including key fire protection insights and factors governing fire risk
- The list of references to support statements and decisions made during the risk assessment
- Inputs to the monitoring process

This documentation is the basis for communicating risk in contributing scenarios, associated fire hazards, and the fire protection features protecting against those hazards. It also serves to establish the monitoring program as key assumptions and fire protection features are identified.

In some applications, the documentation associated with the fire risk assessment may become the basis for fire protection regulatory compliance. As such, the

SFPE Guide to Fire Risk Assessment, The Society of Fire Protection Engineers Series, https://doi.org/10.1007/978-3-031-17700-2_14

document will govern the activities related to the monitoring process so that the following are maintained:

- Design inputs and assumptions to ensure the conclusions remain valid.
- The actions necessary to implement and manage the fire protection program.
- The risk mitigation measures are effective throughout the lifecycle of the building or process.

Chapter 15
Residual Risk Management and Monitoring

Following the implementation of the selected design option, the residual risk needs to be managed, and the risk assessment should be routinely monitored to ensure its applicability. Residual risk management relates to what to do with the risk level resulting from the analysis, accept it as it is, transfer it, or reduce it further. The risk assessment should be routinely monitored to ensure its applicability. This includes a routine check-in that consists of a programmatic approach for:

- Any new applicable codes or standards
- New hazards added to the facility
- Changes in the exposure of people, property, or activities to existing hazards
- Physical modifications
- Changes in analysis inputs, assumptions, and acceptance criteria

Specifically, risk monitoring refers to identifying risk-contributing elements to ensure that the risk estimates are maintained over time as conditions in the building or facility change over time.

15.1 Residual Risk Management

Fire risk will never be zero. Society generally expects that the fire risk associated with the operation of a facility is tolerable and acceptable by the authority having jurisdiction. Residual risk refers to the level of risk that has been accepted for the facility. This residual risk needs and is often managed using one or more of the following three approaches:

- Stakeholders may accept the residual risk. The facility owner/operator will assume the loss associated with any fire once the fire protection program associated with the facility is accepted by the authority having jurisdiction.

SFPE Guide to Fire Risk Assessment, The Society of Fire Protection Engineers Series, https://doi.org/10.1007/978-3-031-17700-2_15

- A facility operator/owner may transfer the residual risk. In this approach, the residual fire risk will be transferred to an insurance company that would assume the loss associated with a fire accident. In this approach, the insurance company may request additional fire protection requirements in addition to those required by the authority having jurisdiction.
- A facility operator/owner may work on reducing the residual risk. This is consistent with the ALARP principle described earlier in Chap. 7. In this approach, the residual risk may be reduced as much as practical by physical modifications to the facility (e.g., installing additional fire protection systems), changes in operational practices (e.g., removing fire hazards), or improvements in fire safety procedures. Although this approach may not fully eliminate the fire risk, it may assist in managing residual risk under the first two options described earlier.

15.2 Monitoring Program Development and Implementation

A monitoring program is implemented to determine what elements are required to maintain the safety/risk at an acceptable level as determined by the fire risk assessment over time. The key features of a monitoring program are discussed below.

15.2.1 Identifying Elements

The first step in developing a risk monitoring program is to identify and list each of the elements included in the fire risk assessment. In practice, the results of uncertainty and sensitivity assessment may be used to identify the key elements in the fire protection program requiring monitoring.

15.2.2 Key Element Risk Ranking

Each credited element should be evaluated within the context of the fire scenarios. This will allow the elements most critical to maintaining an acceptable safety/risk level to be the focus of the monitoring program. These elements may be determined qualitatively or quantitatively.

- Qualitatively, the impact of each element's effect on the fire risk assessment may be determined using a fire protection evaluation, such as engineering judgment. For example,
  - If the fire department's arrival is not preventing the consequence associated with a scenario in time, it may be determined to be a lower ranking.

- If an acceptable level of risk requires the prevention of fire propagation outside an ignition source, the ability to promptly detect a fire may rank high.

- Quantitatively, a sensitivity analysis (see Chap. 13 for guidance on sensitivity analysis) may be performed for each of the identified elements – the failure probability for an element is set to fail (or not available), and the delta risk, or change, from the base risk assessment, is determined. The calculated delta ranks each element. Elements with deltas that change the risk to an unacceptable level are the elements that may require monitoring for future changes.

Ranking of the elements may be done relative to or against predetermined numerical change in the risk results. The ranking helps determine which elements are the most significant risk and ensures those elements focus on the monitoring program.

Elements determined not significant may not need to be monitored. This allows for the focus of and cost of the monitoring program to be dedicated to only the elements necessary to maintain an acceptable level of risk. At the same time, this does not mean that elements that are not monitored are removed or eliminated from the analysis. Elements that may be low-risk contributing should be included in the quantification as the cumulative effects of modifications to multiple elements determined may eventually increase their significance. For example, relatively minor alterations to the spatial configuration or technology may increase the level of risk when they occur repeatedly. In other words, the modification should not be evaluated within the full scope of the risk assessment, including the cumulative effect of any other preceding modifications.

15.2.3 Implementation

Implementation of a monitoring program requires the risk-relevant elements to be evaluated periodically. The frequency at which individual elements are monitored may be determined by type-specific code which requires inspection, testing, and maintenance (ITM) frequency.

Over the life of the building, facility, or process, there may be instances where a risk-relevant element becomes unavailable (e.g., taken out of service for maintenance or has failed and need to be repaired or replaced). In such situations, the unavailability of the impaired element with respect to risk should be evaluated. If the impairment impacts a risk element/scenario, a compensatory may need to be implemented to ensure that an appropriate level of fire safety is maintained.

It is noted that some elements in the monitoring program are easily identified, such as the failure of a fire protection system or the introduction of a new hazard source. However, some changes may occur slowly over time and are not recognized or observed during the monitoring process. Examples could include increased unreliability of a fire protection system or activities that change over time and influence the risk assessment.

Correction to: SFPE Guide to Fire Risk Assessment

Correction to:
Society for Fire Protection Engineers, *SFPE Guide to Fire Risk Assessment*, The Society of Fire Protection Engineers Series, https://doi.org/10.1007/978-3-031-17700-2

The author name was inadvertently displayed with an error in the website. This has now been changed from 'Austin Guerrazzi' to 'Society of Fire Protection Engineers'. The initially published version has now been corrected.

The updated original version of this book can be found at:
https://doi.org/10.1007/978-3-031-17700-2

SFPE Guide to Fire Risk Assessment, The Society of Fire Protection Engineers Series, https://doi.org/10.1007/978-3-031-17700-2_16

Annex: Conceptual Example

This annex describes a conceptual example following the guidance presented in this guide. The example consists of the development of a simplified fire risk assessment for a computing facility. Although representative of some industries, the input values used throughout the examples should not be treated as referenceable material. No technical basis for those inputs has been developed as part of this effort.

A.1. Project Objectives and Scope

For this example, the objective is to determine the risk associated with the facility being unable to operate in the event of a fire in the server room. The scope of the analysis is defined as follows:

- Location and boundaries: While the facility has multiple rooms, the example assessment is limited to fires occurring in the server room.
- Hazards: Fire hazards within the scope of the analysis include those associated with electrical equipment permanently located in the room and those ignition sources and combustibles routinely brought into the room temporarily as part of work-related activities or arson.
- Fire protection elements: The assessment considers the impact of:
  - Automatic fire protection features identified in the server room
  - Internal fire response procedures governing the operation of the room
  - Local fire department response to a fire event in the room
- Affected parties: The assessment is only limited to any impact on business interruption of the computing facility. Assessment of the possible loss of life at the facility is not assessed.
- Type of assessment: The assessment is performed both qualitatively and quantitatively.

SFPE Guide to Fire Risk Assessment, The Society of Fire Protection Engineers Series, https://doi.org/10.1007/978-3-031-17700-2

A.2 Design Information and Data Collection

Chapter 5 identifies information that should be collected to support the development of a fire risk assessment. The following data has been collected to support the fire risk assessment:

- General layout drawings of the facility, including the server room identifying:
  - Wall materials
  - Door locations
  - Locations of automatic smoke detection devices
  - Barrier penetrations such as doorways and HVAC vents
- Fire incident and experimental data (for example):
  - The frequency of fires in the types of cabinets observed during walk-downs
  - Experimental fire testing of fires occurring in the type of equipment observed in the server room during walk-downs
  - Reliability and availability values for the automatic smoke detection devices
- Design specifications including procedures for:
  - Combustible storage
  - Housekeeping (e.g., waste removal)
  - Security access
  - Fire response by on-site security personnel
- Occupant information as described in the design specifications and procedures
- Process documentation and drawings, including the design specifications for the equipment located within the server room.
- Walk-downs were performed to collect the information necessary to develop fire scenarios, including:
  - Fire hazards present in the room (e.g., ignition sources, secondary combustibles)
  - Fire detection and suppression features
  - Room construction and layout

The necessary information listed above can be practically classified into three groups:

(a) Information collected from engineering, architectural, or design drawings
(b) Information collected during walk-downs
(c) "Generic" information for supporting input parameters such as ignition frequencies, fire protection system availabilities.

The first two classifications listed above are, in most cases, project-specific and relatively easy to obtain. Difficulties in getting this information are associated with accessibility to facility information and to the facility itself. Consequently, this example assumes the information is available. In contrast, generic data to support key input parameters to the fire risk assessment (e.g., ignition frequencies) that are usually not part of the design information or collected through walk-downs may require research. The following subsections provide guidance on this topic focusing on this example. These subsections should serve as general guidance and do not constitute comprehensive research on these topics.

A.2.1 Information for Supporting Fire Ignition Frequencies

The fire risk assessment requires enough information to determine ignition frequencies. Depending on the information available, the assessed values may need to be characterized as uncertain. The impact of this uncertainty will eventually need to be evaluated through sensitivity and uncertainty analysis. In this specific application, readily or publicly available generic data supporting the calculation of ignition frequencies will likely not be available.

A limited and informal Internet search suggests that data center fires or server room fires often go unreported to minimize insurer response, limit public knowledge, and other reasons. Although fires may not be the most common causes of downtime, they do occur and can cause major service interruptions. Consequently, quantifying frequencies will require an uncertainty characterization due to the low reportability of ignition events. In addition, these considerations which affect the assessment of ignition frequencies were identified:

- The number of server rooms or data center facilities have been growing over time.
- The size and design of server rooms or data centers vary widely.
- It is likely that only relatively large fires are reported. Therefore, the number of ignition events, particularly those associated with small fires, are underreported or never reported.
- Even if the specific facility has good record-keeping of fire events, the available data may not represent or capture generic ignition frequency trends due to the relatively large universe of similar facilities.

The limited research performed for this example included a review of news articles describing fire events, blog commentary between IT professionals related to fires in data centers and server rooms, fire protection recommendations for these facilities from suppliers, and HVAC recommendations for these facilities from suppliers. Based on this research, it is concluded that:

- Ignition events not developing in relatively large fires do occur. Examples include fires in power supply cabinets.
- Numerous facilities have operated for years without experiencing an ignition event.
- Leading causes for fires include equipment heat up, electrical faults in electrical equipment, dust buildup, cables overheating, and fires resulting during maintenance operations.

- There are fires that have developed to damage entire rooms and further propagated to adjacent spaces.
- Given the potential consequences of fire events in terms of downtime, reliance on fire protection strategies is common.

A.2.2 Information for Supporting Hazards Analysis

The limited research performed for this example included a review of news articles describing fire events, blog commentary between IT professionals related to fires in data centers and server rooms, fire protection recommendations for these facilities from suppliers, and HVAC recommendations for these facilities from suppliers. Based on this research, the leading causes for fires include equipment heat up (e.g., due to loss of room cooling), faults in electrical equipment, dust buildup, cables overheating (e.g., cable bundles in under floors), and fires resulting during maintenance operations.

A.2.3 Information for Supporting System Reliability and Availability

A similar limited and informal Internet review was performed to assess the reliability of smoke detection systems. The information available suggests that the systems are often better than 70% reliable (i.e., failure probability after a mission time of 30%). This value is generic and likely conservative as it includes different types of facilities, occupancies, and inspection, testing, and maintenance practices. For example, a value of 95% is recommended as a generic reliability value to be used in fire risk assessments for commercial nuclear facilities (see Appendix P of NUREG/CR-6850 Vol 2 [1]).

A.3 Risk Assessment Method Selection

This example is solved using both the qualitative and the quantitative methodology described in this guide.

A.4 Acceptance or Tolerance Criteria

The acceptance or tolerance criteria used in this example are shown in the qualitative and quantitative risk matrix described earlier in Chap. 7 (see Tables 7.4 and 7.5).

A.5 Hazard Identification

Recall that hazards can be identified by reviewing fire events at similar facilities, a walk-down of the facility, and engineering judgment as identified in Sect. 8.4. The hazard classifications identified in Table 8.1 can also assist in the process. For this example, a walk-down of the server room and interviews with facility operators identified the following information:

- The storage of combustible material is procedurally prohibited within the server room. The door to the room is marked with a sign stating that storage of any material is prohibited within the server room. A similar sign is located within the server room.
- Housekeeping (e.g., trash removal) is performed daily.
- Key card access is required to enter the server room. Each employee has an ID card that allows them access to the server room.
- All the furniture within the server room is noncombustible.
- Cables – control and power – are all routed to limit the opportunity for fraying or damage. Most cables are below a false floor. All cables located above the floor are routed in metal conduits when outside a cabinet.
- There are three types of equipment within the server room: a power supply cabinet, a control cabinet, and a server bank.
- There are two automatic spot smoke detection devices within the room. There is no automatic suppression system.
- The walls of the server room are solid concrete. There is a false floor and ceiling; both are constructed of noncombustible material.

Reviewing this information, many hazard classifications presented in Table 8.1 are identified and reviewed in Tables A.1, A.2, and A.3 following a "What if" analysis. The "what if" analysis is presented in this example for completeness purposes in the process of developing a fire risk assessment. This example's objective is not to provide a detailed description of the "What if" process as a hazard analysis tool.

Table A.1 What if hazard identification example: combustible material

Location: server room		Description: combustible material in the server room		By: review team
What if	Answer	Consequences	Recommendations	Discussion
General storage is allowed	Increases combustible loading in the space (i.e., additional secondary combustibles or ignition sources)	Consequences resulting from igniting larger quantities of secondary combustibles will be defined later as part of the definition and characterization of fire scenarios	Prohibit storage of combustible material in the server room	This element is an example of a propagation hazard as combustible material storage may provide additional fuel following ignition and increase the consequences associated with a fire in the server room. It can also be an example of additional ignition sources in the room for which fire scenarios may need to be defined During walk-downs, no storage of combustible material was observed
No housekeeping is performed	Increases combustible loading in the space (i.e., additional secondary combustibles or ignition sources) Dust buildup in electrical equipment or cable runs	Consequences resulting from igniting larger quantities of secondary combustibles will be defined later as part of the definition and characterization of fire scenarios	Enact and follow housekeeping procedures for removing unnecessary combustible materials	This is an example of a possible human-caused precursor hazard This may not represent an ignition source but could influence the likelihood of the presence of combustible material Procedures applicable to the computer room note that housekeeping is performed daily
Combustible furniture is used	Increases combustible loading in the space	Consequences resulting from igniting larger quantities of secondary combustibles will be defined later as part of the definition and characterization of fire scenarios	Ensure all furniture used in the room is constructed from noncombustible material	This is an example of an equipment propagation hazard The walk-through identified that all furniture in the room is noncombustible. Therefore, no increase, in consequence, needs to be considered due to the presence of furniture in the server room
Combustible dust or solids buildup within the equipment	Increases combustible loading in the space	Consequences resulting from igniting larger quantities of secondary combustibles will be defined later as part of the definition and characterization of fire scenarios	Include cleaning of dust into regular equipment inspection, testing, and maintenance (ITM) practices	This is an example of a process propagation hazard While there are procedures for daily housekeeping, it may not be assumed that general housekeeping involves steps to ensure the buildup of dust in hard-to-reach places such as within cabinets or below the false floor are included. Therefore, this may need to be considered as an element that could lead to an increased risk

Table A.2 What if hazard identification example: ignition source

Location: server room		Description: ignition sources		By: review team
What if	Answer	Consequences	Recommendations	Discussion
Equipment overheats (e.g., loss of room cooling)	Ignition source	Consequences resulting from ignition will be defined later as part of the definition and characterization of fire scenarios	Perform regular inspections of equipment conditions. Maintain equipment within their recommended operating range. Establish compensatory measures to address periods where room cooling is not available	This element is an example of an equipment ignition hazard The two cabinets and server bank should be considered as possible equipment ignition sources
Maintenance activities produce a fire	Maintenance activities can produce ignition that could further propagate through cables or other combustibles	Consequences resulting from igniting combustibles will be defined later as part of the definition and characterization of fire scenarios	Provide training on good practices while performing maintenance work	This is an example of a possible human-caused precursor hazard This may not represent an ignition source but could influence the likelihood of the presence of combustible material Procedures applicable to the computer room note that housekeeping is performed daily
Cables run damaged (e.g., underfloor cable runs overheats)	Ignition source	Consequences resulting from ignition will be defined later as part of the definition and characterization of fire scenarios	Perform regular inspections of cable conditions	This element is an example of a human precursor hazard as inspections of cable conditions was not performed. This could also be a hazard associated with cable fires. A suppression response may be delayed for a fire in the cables routed below the false floor
Electrical equipment fails (e.g., electrical faults)	Ignition source	Consequences resulting from ignition will be defined later as part of the definition and characterization of fire scenarios	Perform regular inspections of cable conditions	This element is an example of an equipment ignition hazard The two cabinets and server bank should be considered as possible equipment ignition sources
Arson	Ignition source	Consequences resulting from ignition will be defined later as part of the definition and characterization of fire scenarios	Control access to the server room	This element is an example of a human precursor hazard Control access to the server room may limit the possibility of an arson event. However, control access may delay a suppression response and increase the consequence of a fire in the server room due to non-arson ignition events

Table A.3 What if hazard identification example: response to alarm

Location: server room		Description: response to alarm		By: review team
What if	Answer	Consequences	Recommendations	Discussion
Responders: First responders unable to enter space	Delayed suppression response	Consequences for delayed suppression activities will be defined later as part of the definition and characterization of fire scenarios	Provide training for first responders	NOTE: Failure of fire protection features to operate is not a hazard triggering a fire scenario. Instead, the failure of these systems is often explicitly captured in the risk assessment through the use of conditional probabilities
Smoke detection system does not operate	Delayed suppression response	Consequences for delayed suppression activities will be defined later as part of the definition and characterization of fire scenarios	Perform regular inspection, testing, and maintenance for the fire detection system	NOTE: Failure of fire protection features to operate is not a hazard triggering a fire scenario. Instead, the failure of these systems is often explicitly captured in the risk assessment through the use of conditional probabilities

In summary, the most common fire hazards in data centers or server rooms are associated with equipment failures (e.g., electrical failures in power supply cabinets), problems with underfloor wiring, housekeeping (e.g., combustible controls, dust buildup), equipment overheating due to loss of room cooling or fires initiated by maintenance activities.

A.6 Fire Scenarios

The hazards identified in the previous section are used as the starting point for defining and characterizing the fire scenarios necessary to characterize the risk appropriately. Recall that the following set of elements is used to characterize fire scenarios: frequency, location, ignition source, intervening combustibles, fire protection features, and consequences. Table A.4 lists the identified scenarios and the hazards they cover.

Table A.4 Computer facility fire scenarios

Scenario #	Hazard	Notes
Scenario 1	Arson	Security procedures may limit the likelihood of this hazard
Scenario 2	Storage of materials/ combustible furniture	Table A.1 identifies that general storage and the use of combustible furniture could serve as combustible materials within the server room. While not observed during walk-downs, these may represent human-caused issues The desk observed during walk-downs is constructed of noncombustible materials
Scenario 3	Trash (transient combustible)	Section A.5 noted that procedures are in place for daily housekeeping space; however, the presence of trash represents a non-fixed (transient) source of combustible material
Scenario 4	Server bank	Table A.2 Notes the possibility of equipment overheating as an ignition source hazard
Scenario 5	Control cabinet	Table A.2 Notes the possibility of equipment overheating as an ignition source hazard
Scenario 6	Power supply cabinet	Table A.2 Notes the possibility of equipment overheating as an ignition source hazard

This example elaborates on a single scenario only: the power supply cabinet scenario. It is noted that a full risk assessment will require expanding the analysis to all the identified scenarios.

From the information gathered through the data collection process, the cabinet is a typical design used frequently in these types of facilities and has no history of failures generating fires. The power supply cabinet is located within a compartment (room) protected by automatic spot smoke detection devices. There is no fixed automatic or manually operated suppression system. The detection system sends an alarm signal to a monitored security station. By procedure, the response of the personnel stationed in the security station is expected to be less than 5 min following the actuation of an alarm. The personnel are trained in the use of fire extinguishers. If necessary, an off-site response by the fire department is expected within 15 min following activation of the automatic detection system.

Walk-downs suggest that while there are no intervening combustibles expected to ignite by a fire in the power supply cabinet, there are at least two other components that a fire could damage: a control cabinet located directly adjacent to the power supply cabinet and a server bank located nearby. A layout of the server room, as sketched during the walk-downs, is presented in Fig. A.1.

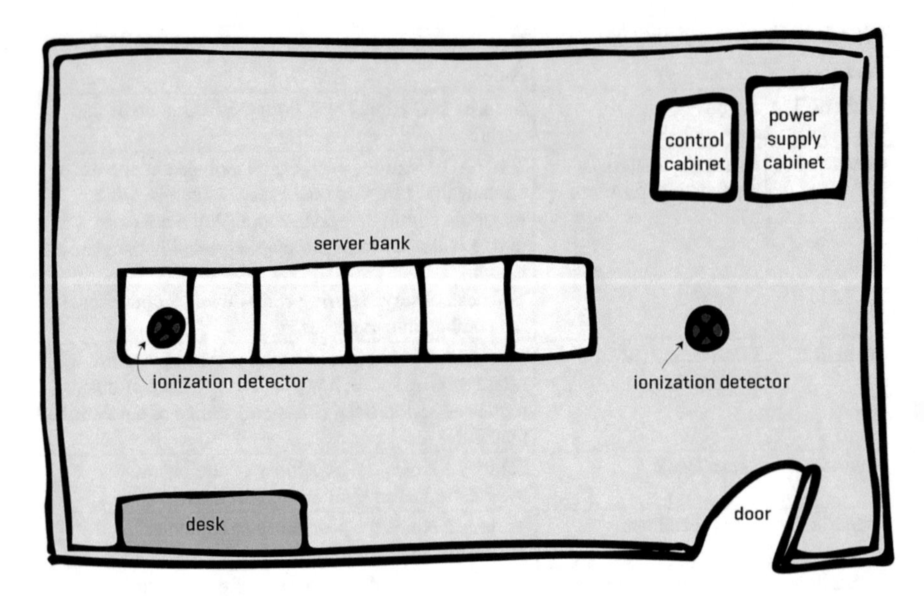

Fig. A.1 Example: server room

Discussions with the facility operators determined that there is limited risk to the facility's operations due to the loss of the power supply cabinet, as its functions are supported by a redundant power supply outside the room. The failure of the control cabinet has the potential to interrupt the operations of the facility. In addition, damage to the server bank would likely result in interruptions of the facility's operations. While there are procedures to limit the impact of damage to this equipment, its loss due to fire will still result in business interruption.

Characterization of Scenario 6:

- *Ignition Source*: Power supply cabinet.
- *Frequency*: Since the source cabinet is a typical design, there may be fire incident data to determine ignition frequency.
- *Location*: The source cabinet is located within a compartment. Location may influence what secondary combustibles could contribute to the fire or the presence of personnel who could detect or suppress a fire.
- *Intervening Combustibles*: There are no items external to the source cabinet that could propagate to and influence the size or power of the fire considered.
- *Fire Protection Features*: The compartment is protected by automatic spot smoke detection devices. There is an expected manual response within 5 min and a more substantial fire department response expected within 15 min. The fire compartment is constructed of fire-rated materials.
- Consequences: The evaluation of consequence is business continuity: specifically, the loss or interruption of facility functions. The loss of each of the three equipment components is reviewed:

- Power supply cabinet: Redundant cabinet, limited or no risk to facility operation.
- Control cabinet: The loss of this cabinet has the potential to affect facility operations.
- Server bank: The loss of this equipment would result in a significant loss of function at the facility.

From the information above, the number of scenarios to analyze is determined. The number of damage states (See Fig. 9.1) and the subsequent number of scenarios may be determined by the increasing progression of consequences. In this scenario, the risk associated with fire propagation to four damage states are evaluated:

- Damage State 0: Damage limited to the ignition source.
- Damage State 1: Damage to the control cabinet (located directly adjacent to the ignition source).
- Damage State 2: Damage to the bank of servers located nearby the ignition source.
- Damage State 3: Damage to the entire room due to the development of a damaging smoke layer.

Finally, the qualification of the suppression activities for each damage state is necessary to perform the risk quantification. For this example, the qualification of successful suppression responses for each damage state is determined by a combination of detailed fire modeling and the detection information provided by the facility. These estimates are as follows:

- Damage to targets considered in Damage State 1 occurs within 1–5 min.
- Damage to targets considered in Damage State 2 occurs within 10 min.
- Damage to targets considered in Damage State 3 is not expected to occur.

A.7 Qualitative Fire Risk Estimation and Evaluation

An event tree approach similar to the one performed in Chap. 10 will be used to qualitatively assess the fire risk associated with the power supply cabinet. The event tree for this example is presented in Fig. A.2.

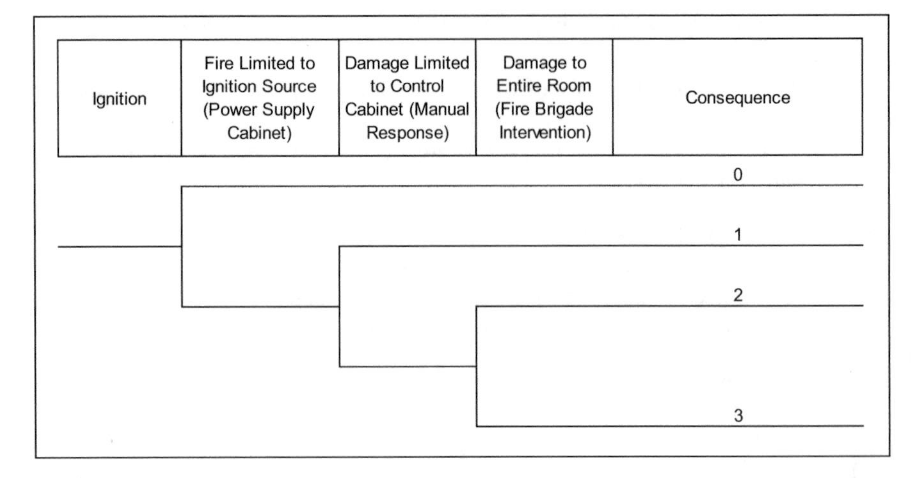

Fig. A.2 Scenario progression event tree representing the cabinet fire

The events are:

- *Ignition*: This is the initiating event. It is noted that the initiating event captures all the possible failure modes of the power supply cabinet leading to ignition (e.g., equipment overheats, electrical faults). A review of the cabinet determined that it is of a typical design. As discussed earlier in this Annex, there is evidence of fires in this type of cabinet in commercial buildings. Therefore, an "Occasional" frequency level is assigned given the research performed and the evidence collected. Recall that an "occasional" level suggests that a fire is likely to occur sometime in the lifetime of an item or will happen several times in the operation of a large number of similar items.
- *Fire Limited to Ignition Source (Power Supply Cabinet)*: Using the FSCT tree to support the analysis, the *Manage Fire* and *Manage Exposed* (under the Manage Fire Impact gate) branches may be used to systematically assess the elements that influence the fire's ability to propagate beyond the ignition source. The closest item of interest in the scenario progression is the control cabinet. Exploring the FSCT, there is a limited impact to be expected following the *Manage Exposed* branch:
  - The *Move exposed* branch is not applicable (classified as *Nonexistent* in the FSCT) as both the power supply and control cabinet are fixed locations. For this example, moving the cabinet is not a practical option given the as-built conditions of the room.
  - The *Defend exposed in place* branch is also not applicable at this stage as the first manual response would not be expected to arrive in time, and there are no noncombustible barriers between the cabinets.
  - Finally, being located directly adjacent to the power supply cabinet, no credit is given to the *Limit amount exposed* branch in this example.

Under the Manage Fire branch, at this stage in the example, the *Confine/contain fire* event under the *Control fire by construction* branch as the solid construction of the ignition source cabinet would limit the initial thermal damage to the control cabinet. Under the Control combustion process branch, the *Limit fuel quantity*, *Control fuel distribution*, and *Control physical properties of the environment* may be considered "Standard" as the construction of the power supply cabinet follows applicable electrical standards.
The qualitative evaluation classifies the fire safety objective in the FCST as "Standard" for this event. The practical implication of this qualitative assessment is that the failure branch of this event leads to Scenarios 2 through Scenario 4 (i.e., Consequence 2 through Consequence 4) should have a low likelihood associated with the failure of a fire prevention capability that has been evaluated as "standard." Therefore, most of the probability will be apportioned to Scenario 1 (Consequence 1″), representing fires limited to the ignition source with no propagation.

- *Manual Response*: It is assumed that the manual response could detect and control the fire before the automatic smoke detection system alarms. However, it is also assumed for this example that a review of the detection system in the server room was evaluated to be "below standard" (based on the classifications described in NFPA 550) with respect to the ability to detect the hazard presented by the power supply cabinet on a timely matter. This is due primarily to staff not being routinely present in the room. Although procedures and training records for the manual response are available, a prompt response by nearby personnel should not be available continuously.
- *Automatic Detection*: The automatic detection system is evaluated as "standard." In practice, this reflects the code compliance, effectiveness, and availability of the system. Although not explicitly included in the event tree depicted in Fig. A.2, it will be implicitly captured in the ability of the fire brigade to respond in a timely matter.
- *Fire Brigade Intervention*: The facility maintains a fire brigade following state-of-the-art practices and code requirements. In addition, the smoke detection system, which is evaluated as standard, will provide an alarm necessary to trigger a response. Therefore, the Fire Brigade Intervention scenario is ranked as "Standard." Recall that this classification should consider the brigade's ability to control the fire before the postulated consequences are achieved. A time to damage calculation using fire modeling may be necessary to support a conclusion of a timely response. Such fire modeling calculations are outside the scope of this guide. For this example, it is assumed that the calculation is available and, together with training records from the fire brigade that suggest a timely response is likely.

The following consequence levels are assigned:

- *Consequence level 0 (Fire Scenario 0):* A fire in the power supply cabinet controlled by fire prevention measures and does not propagate to nearby combustibles has negligible consequences. This represents damage to the cabinet that requires simple part replacement or replacement of the power supply cabinet. Given the availability of backup power, no significant business interruption is expected.

- *Consequence level 1 (Fire Scenario 1):* The personnel response may not successfully respond to the detected fire. A delayed response is expected to have marginal consequences before involving the control cabinet. This is because the consequences are still mostly limited to the power supply cabinet.
- *Consequence level 2 (Fire Scenario 2):* A fire that grows to an intensity where the fire watch is no longer capable of controlling it and, at the same time, propagation to the control cabinet may produce major consequences. There is no automatic suppression system available to mitigate this consequence.
- *Consequence level 3 (Fire Scenario 3):* Associated with a scenario requiring the fire brigade or fire department to control the fire before room-wide damage is generated. A fire resulting in room-wide damage conditions can be associated with consequences ranging from critical to catastrophic.

Figure A.3 depicts the qualitative assessment in the scenario progression event tree. It is noted that a "standard" classification has the practical effect of maintaining the frequency of the scenario for the damage state as most of the frequency will be associated with the top branch of the event split. The bottom branch of the event split captures the effect of the failed fire protection feature by lowering the frequency at that point in time in the scenario progression. In contrast, a "below standard" classification maintains the frequency of the event at that point in the scenario progression as the fire protection feature is assessed not to impact the scenario progression at that time. The sensitivity of the manual response classification is presented in Section A.9.3.

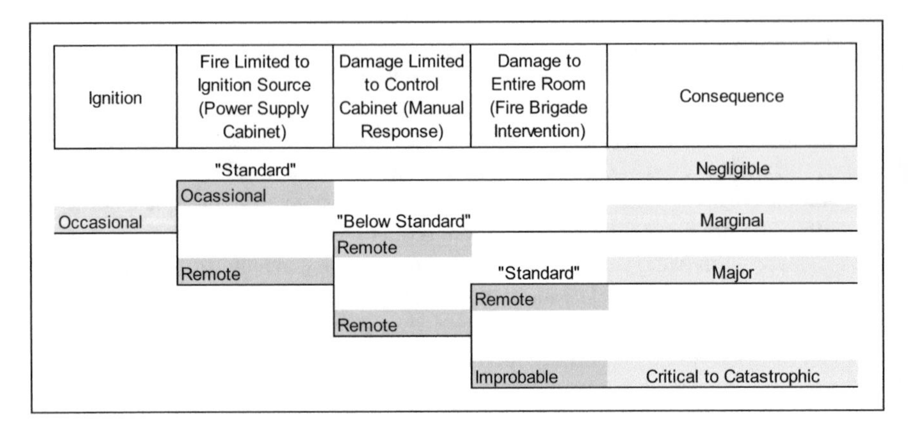

Fig. A.3 Event tree for power supply cabinet example with assessment for consequences

Fire Scenario 0 is associated with negligible consequences. The ignition frequency is "occasional." Due to the negligible consequences, this outcome is associated with acceptable risk, and no further evaluation is necessary. It is noted that fire prevention measures are explicitly credited in the analysis and should be monitored throughout the facility's operation.

Fire Scenario 1 is associated with marginal consequences. At the same time, the scenario frequency consists of a "remote" scenario frequency. In this example, these capabilities have been evaluated as "below standard," which has the practical effect of not reducing the scenario frequency. Therefore, Scenario 1 results in a remote scenario frequency with marginal consequences, which is acceptable.

Scenarios 2 and 3 are associated with major to potentially catastrophic consequences. These scenarios result from the reliance on the fire brigade or fire department. Since the response is likely to be on time, the scenario frequency is not further reduced from "remote" for the successful brigade suppression branch. At the same time, the failure of the fire brigade is represented with an improbable classification to capture the capability in the outcome of the scenario. These scenarios, with remote/improbable and major/catastrophic risk, are acceptable based on the risk matrix presented in Table A.4. It is noted that the catastrophic potential is associated with the failure of the fire brigade or fire department to control the fire on time. It is also noted that relatively large fires starting in data centers and server rooms damaging large sections of a building and generating significant downtime can occur. This is a strong suggestion that the performance of the fire brigade needs to be monitored (including training, etc.) throughout the operational life of the facility.

The risk matrix suggests further evaluation for these scenarios as the "remote" nature of the event, given the fire watch already provides most of the risk reduction. Depending on the level of consequences, additional fire protection features may be necessary. At the same time, it may be practical to accept the residual risk as it is small and the existing fire protection features included in the analysis will be required to be maintained and monitored. Table A.5 summarizes the risk evaluation results.

Table A.5 Summary of risk estimation results for hot work example

Scenario	Ignition frequency (initiating event)	Scenario frequency	Consequences	Risk evaluation	Comment
0	Occasional	Occasional	Negligible	Acceptable	Fire prevention measures are effective propagation. Cabinet construction is credited as part of the analysis in preventing fire propagation
1	Occasional	Remote	Marginal	Acceptable	Prompt facility personnel response is not likely
2	Occasional	Remote	Major	Acceptable	Fire brigade or fire department is effective for suppression. This includes the consideration for the automatic smoke detection system
3	Occasional	Improbable	Critical-catastrophic	Acceptable	Low likelihood of fire brigade failure to control the fire. Fire brigade performance needs to be monitored and maintained at the credited level

A.8 Quantitative Fire Risk Estimation and Evaluation

The process starts with characterizing the ignition frequency. The research described earlier in this example suggested an ignition frequency level of occasional. Numerically, the frequency value should be in the order of 0.01–0.1 (an ignition event in the power supply cabinet every 10–100 years). Given the uncertainty associated with the ignition frequency, the potential range of values will need to be evaluated to assess its impact on risk.

The next event in the scenario progression event tree refers to fires limited to the ignition source (i.e., the power supply cabinet). This event may capture two elements in the scenario: the fire prevention practices and the geometric configuration.

The fire prevention practices may be represented using the fraction of times housekeeping has not been performed as required. In this example, it is assumed that the housekeeping was not performed 5% of the time.

The geometric configuration can be numerically reflected using the concept of fire severity, represented earlier in Fig. 11.4. The power supply cabinet is characterized by a gamma probability distribution with parameters $\alpha = 0.52$ and $\beta = 73$. This is the distribution for electrical cabinets that are closed (Classification Group 4b), as discussed in Table 4.2 in NUREG-2178, Vol. 1 [2].

Based on material flammability properties, it is assessed that the ordinary combustibles may ignite if exposed over time to a heat flux of 10 kW/m^2. A fire modeling analyst determines that a 40 kW fire at the cabinet can generate flame radiation levels of 10 kW/m^2 0.3 m away. Therefore, the fraction of fires that can produce 10 kW/m^2 or more is the area under the curve of the gamma probability distribution to the right of 40 kW, or what is equivalent, the probability of $\dot{Q}$ > 40 kW, where $\dot{Q}$ is the random variable for the heat release rate. This can be solved numerically with Microsoft Excel using the gamma probability distribution function as:

$$= 1 - \text{GAMMADIST}(40, 0.52, 73, \text{TRUE}) = 0.3,$$

which solves for the area under the distribution to the right of 40 kW. Combining the two values, the probability used in the scenario progression event tree is $0.05 \times 0.3 = 0.015$.

The next event in the scenario development event tree is associated with facility personnel controlling the fire. The technical approach described in the hot work example earlier in Chap. 11 is used to estimate the probability of facility personnel intervention controlling the fire. A mission time (i.e., a time to damage) is necessary to calculate a non-suppression. This time is often determined using analytical fire modeling tools. In this example, the mission time represents the time to ignition of secondary combustibles. Secondary combustibles could be the adjacent electrical cabinet, ordinary transient combustibles stored near the ignition source, etc. Since a time to damage calculation is outside the scope of this guide, this example assumes a 10-min ignition time. Borrowing time to suppression statistics from fire events in the commercial nuclear industry, the mean suppression rate for electrical fires in Table 5.3 of NUREG 2169 [3] is $\lambda = 0.102$. It is noted that this value was developed using response time to fire events in the commercial nuclear industry exclusively and may not apply to other facilities or occupancies. Factors governing this value include the presence (or nearby presence) of trained facility personnel, procedures for rapid response to fire detection or equipment malfunction signals, the availability of an industrial on-site fire brigade, and strong awareness of fire risks. Using this value, the resulting non-suppression probability for this time step is:

$$\Pr(T > t) = e^{-\lambda t} = e^{-0.102 \cdot 10} = 0.36$$

In some applications, the detection time is included in the calculation depending on what triggers the manual suppression response. If, for example, an automatic smoke detection signal is received in 5 min after ignition, the resulting values are:

$$\Pr(T > t) = e^{-\lambda t} = e^{-0.102 \cdot (10-5)} = 0.6$$

where $10 - 5 = 5$ min, is the time available for suppression after detection. A similar calculation can be performed to determine the probability of fire brigade or fire department failure at 20 min. In this calculation, 20 min is assumed as the time to "room wide" damage, which is also a value that can be determined using fire modeling tools.

$$\Pr(T > t) = e^{-\lambda t} = e^{-0.102 \cdot (20-5)} = 0.22$$

The following table lists the inputs to the event tree calculating the non-suppression probabilities.

Table A.6 Inputs to the event tree calculating the non-suppression probability assuming a time to automatic detection of 5 min

Fire protection capabilities	Include in the scenario	Activation time (min)	Availability
Detection			
Automatic smoke detection	TRUE	5	0.7
Automatic heat detection	FALSE	0	0
Prompt detection (personnel, fire watch)	FALSE	0	0
Suppression			
Prompt suppression (personnel, fire watch)	FALSE	0	0
Automatic sprinklers	FALSE	0	0
Automatic halon	FALSE	0	0
Automatic CO_2	FALSE	0	0
Manually activated system	FALSE	0	0
Facility personnel response	TRUE	0	See event tree

With the inputs listed in Table A.6, the resulting non-suppression probability is 0.6, as depicted in Fig. A.4. This value is calculated for a mission time of 10 min.

	Prompt Actions		Automatic Actions		Manual Actions				
Ignition	Detection	Suppression	Auto Det	Auto Supp	Manual Fixed Supp	Department/Fire Bridage	Sequence ID	Sequence Probability	Outcome
1	0.0E+00	0.0E+00					A	0.0E+00	Suppression
		1.0E+00		0.0E+00			B	0.0E+00	Suppression
				1.0E+00	0.0E+00	4.0E-01	C	0.0E+00	Suppression
						6.0E-01	D	0.0E+00	No Suppression
					1.0E+00	4.0E-01	E	0.0E+00	Suppression
						6.0E-01	F	0.0E+00	No Suppression
	1.0E+00		7.0E-01	0.0E+00			G	0.0E+00	Suppression
				1.0E+00	0.0E+00	4.0E-01	H	0.0E+00	Suppression
						6.0E-01	I	0.0E+00	No Suppression
					1.0E+00	4.0E-01	J	2.8E-01	Suppression
						6.0E-01	K	4.2E-01	No Suppression
			3.0E-01	0.0E+00			L	0.0E+00	Suppression
				1.0E+00	0.0E+00	4.0E-01	M	0.0E+00	Suppression
						6.0E-01	N	0.0E+00	No Suppression
					1.0E+00	4.0E-01	O	1.2E-01	Suppression
						6.0E-01	P	1.8E-01	No Suppression
						Non Suppression Probability		6.0E-01	

Fig. A.4 Detection/suppression event tree solved at 10 min

Similarly, a non-suppression probability is also calculated for the mission time of 20 min, assuming a 5 min automatic detection time. This calculation is described in Fig. A.5.

	Prompt Actions		Automatic Actions		Manual Actions				
Ignition	Detection	Suppression	Auto Det	Auto Supp	Manual Fixed Supp	Department/Fire Bridage	Sequence ID	Sequence Probability	Outcome
1	0.0E+00	0.0E+00					A	0.0E+00	Suppression
		1.0E+00		0.0E+00			B	0.0E+00	Suppression
				1.0E+00	0.0E+00	7.8E-01	C	0.0E+00	Suppression
						2.2E-01	D	0.0E+00	No Suppression
					1.0E+00	7.8E-01	E	0.0E+00	Suppression
						2.2E-01	F	0.0E+00	No Suppression
	1.0E+00		7.0E-01	0.0E+00			G	0.0E+00	Suppression
				1.0E+00	0.0E+00	7.8E-01	H	0.0E+00	Suppression
						2.2E-01	I	0.0E+00	No Suppression
					1.0E+00	7.8E-01	J	5.5E-01	Suppression
						2.2E-01	K	1.5E-01	No Suppression
			3.0E-01	0.0E+00			L	0.0E+00	Suppression
				1.0E+00	0.0E+00	7.8E-01	M	0.0E+00	Suppression
						2.2E-01	N	0.0E+00	No Suppression
					1.0E+00	7.8E-01	O	2.4E-01	Suppression
						2.2E-01	P	6.5E-02	No Suppression
						Non Suppression Probability		2.2E-01	

Fig. A.5 Detection/suppression event tree solved at 20 min

The conditional non-suppression probabilities are presented in Fig. A.6.

Scenario	Elapsed time after fire ignition (minutes)	Estimated non-suppression probability, P_{ns}	Cumulative suppression probability, $P_s = 1 - P_{ns}$	Formula for the conditional probability of the scenario	Conditional probability of the scenario
1	10	0.6	0.4	$P_s(10)$	0.4
2	20	0.22	0.78	$P_s(20) - P_s(10)$	0.38
3	>20			*Remaining Probability*	0.22

Fig. A.6 Summary of the non-suppression probabilities apportioned to each scenario

The non-suppression probabilities listed in the table above are used to calculate the frequency of each scenario sequence resulting from ignition as illustrated in Fig. A.7.

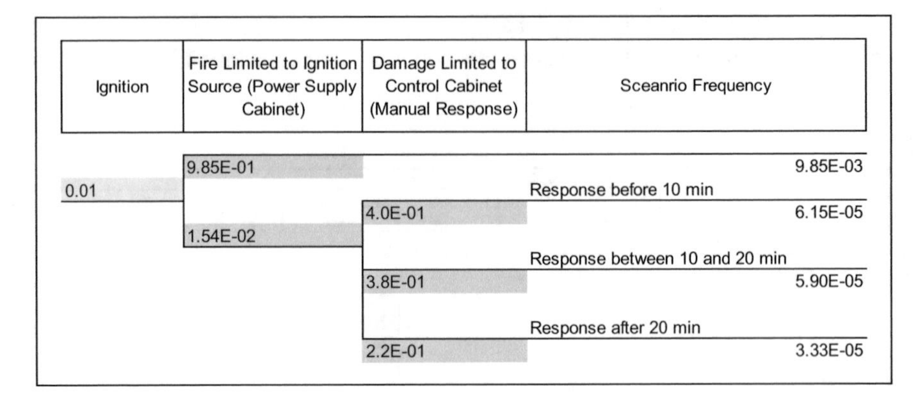

Fig. A.7 Scenario frequencies considering detection and suppression for each event in the chronology

The consequence level can now be assigned to each scenario sequence to determine a risk value. The values in this example are calculated using normalized consequences as described in Chap. 7. These normalized values can also be represented in monetary terms so that the risk metric is monetary losses per ignition (Fig. A.8).

Ignition	Fire Limited to Ignition Source (Power Supply Cabinet)	Damage Limited to Control Cabinet (Manual Response)	Sceanrio Frequency	Consequences		Risk
0.01	9.85E-01		9.85E-03	1.00E-05	Negligible	9.85E-08
	1.54E-02	4.0E-01	6.15E-05	1.00E-04	Marginal	6.15E-09
		3.8E-01	5.90E-05	1.00E-02	Major	5.90E-07
		2.2E-01	3.33E-05	1.00E+00	Critical to Catastrophic	3.33E-05

Fig. A.8 Scenario progression event tree for power supply cabinet example with assessment for consequences

The first fire scenario is associated with negligible consequences. The ignition frequency is "occasional," which is quantified with a value of 1.0E-2/year. Due to the negligible consequences, this outcome is associated with an acceptable risk (i.e., in the order of 1.0E-7/year), and no further evaluation is necessary. It is noted that fire prevention measures are explicitly credited in the analysis and should be monitored throughout the facility's operation. In this scenario, the risk assessment suggests that fire prevention measures are one of the most influencing factors in the results.

Fire Scenario 1 is associated with marginal consequences. At the same time, the scenario frequency consists of a "remote" scenario frequency (6.15E-5/year). As mentioned earlier, this value is heavily influenced by fire prevention practices as the probability of controlling or suppressing the fire at this point is relatively low. Therefore, Scenario 1 results in a remote scenario frequency with marginal consequences, which is acceptable.

Scenarios 2 and 3 are associated with major to potentially catastrophic consequences. These scenarios result from the reliance on the fire brigade or fire department. The quantification suggests that the sequence with the largest risk contribution is the one associated with catastrophic consequences. These scenarios with remote and major/catastrophic risk have an acceptable risk based on the risk matrix presented in Table 7.4. However, the risk matrix suggests that a risk level in the order of 1E-4/year to 1E-5/year may require further evaluation. Since the catastrophic potential (i.e., relatively large fires starting data centers or server rooms damaging large sections of a building and generating significant downtime have occurred) are driving the risk results, higher failure probabilities for the fire brigade's or fire department's ability to control the fire on time and failure of fire prevention measures can significantly increase the risk to non-acceptable levels. Therefore, given the unavailability of automatic fire suppression, these two elements have to be monitored and, if possible, improved over time. Table A.7 summarizes the risk evaluation results.

Table A.7 Summary of risk estimation results for server room example

Scenario	Ignition frequency (initiating event)	Scenario frequency	Consequences	Risk evaluation	Comment
0	0.01	9.85E-3	1.0E-5	9.8E-8	Fire prevention measures are effective propagation. Cabinet construction is credited as part of the analysis in preventing fire propagation
1	0.01	6.15E-5	1.0E-4	6.15E-9	Prompt facility personnel response is not likely, but consequences are low
2	0.01	5.90E-5	1.0E-2	5.9E-7	Fire brigade or fire department is effective for suppression. This includes the consideration for the automatic smoke detection system
3	0.01	3.33E-5	1.0	3.3E-5	Relatively low likelihood of fire brigade failure to control the fire. Fire brigade performance needs to be monitored and maintained at the credited level. An increase in fire brigade failure probability can lead to unacceptable risk levels

Recall that the qualitative and quantitative example developed in this section was limited to only one scenario. The risk associated with the server room or the entire facility will need to include quantifying all the identified scenarios.

A.9 Sensitivity and Uncertainty Analysis

Three elements in the risk assessment process can be identified as either uncertain or heavily influencing the calculated risk values: the ignition frequency, the effectiveness of the fire prevention measures, and the fire brigade's ability to control the fire on time. In addition, the alternative of the installation of an automatic suppression system can be evaluated.

A.9.1 Sensitivity Analysis for the Fire Ignition Frequency

This example characterized the ignition frequency as uncertain due to the challenges associated with obtaining fire events data. Based on the information available, a frequency classification of "occasional" in the order of 1E-2/year was assigned. Recall that this value represents ignition events, including those that may not have resulted in a relatively large fire. Evaluating the sensitivity in the ignition frequency parameter is relatively simple as it is a constant multiplier to the risk numbers. Increasing the ignition frequency by one order of magnitude suggests risk values with catastrophic consequences in the order of 4.0E-4/year. Individually, this scenario may still present an acceptable or tolerable risk level. However, considering that this is only one scenario in the facility, the contribution from other fire scenarios in similar cabinets will suggest unacceptable risk levels requiring increased levels of fire safety. Given the uncertainty associated with this parameter, the fire protection engineer may need to consider strategies to improve fire safety (Table A.8).

Table A.8 Summary of sensitivity analysis results increasing the ignition frequency by one order of magnitude

Scenario	Ignition frequency (base)	Ignition frequency (sensitivity)	Conditional probability	Consequences	Risk evaluation (base)	Risk evaluation (sensitivity)
0	0.01	0.1	9.9E-01	1.0E-05	9.9E-08	9.9E-07
1	0.01	0.1	6.2E-03	1.0E-04	6.2E-09	6.2E-08
2	0.01	0.1	5.9E-03	1.0E-02	5.9E-07	5.9E-06
3	0.01	0.1	3.3E-03	1.0E+00	3.3E-05	3.3E-04

A.9.2 Sensitivity Analysis for Fire Prevention Practices

Figures A.9 and A.10 illustrate the impact of increasing the failure probability of fire prevention measures. The results highlight the importance of this element in the fire protection program, as it lowers the risk. The lack of fire prevention will result in unacceptable risk levels. Such results suggest the importance of monitoring activities such as housekeeping, equipment inspections, and ensuring proper environmental parameters for the operation, identified as key factors that increase the risk of fires in server or data center rooms.

Ignition	Fire Limited to Ignition Source (Power Supply Cabinet)	Damage Limited to Control Cabinet (Manual Response)	Sceanrio Frequency	Consequences		Risk
0.01	9.00E-01		9.00E-03	1.00E-05	Negligible	9.00E-08
	1.00E-01	4.0E-01	4.00E-04	1.00E-04	Marginal	4.00E-08
		3.8E-01	3.83E-04	1.00E-02	Major	3.83E-06
		2.2E-01	2.17E-04	1.00E+00	Critical to Catastrophic	2.17E-04

Fig. A.9 Sensitivity analysis increasing the failure probability of fire prevention to 0.1

Ignition	Fire Limited to Ignition Source (Power Supply Cabinet)	Damage Limited to Control Cabinet (Manual Response)	Sceanrio Frequency	Consequences		Risk
0.01	5.00E-01		5.00E-03	1.00E-05	Negligible	5.00E-08
	5.00E-01	4.0E-01	2.00E-03	1.00E-04	Marginal	2.00E-07
		3.8E-01	1.92E-03	1.00E-02	Major	1.92E-05
		2.2E-01	1.08E-03	1.00E+00	Critical to Catastrophic	1.08E-03

Fig. A.10 Sensitivity analysis increasing the failure probability of fire prevention to 0.5

A.9.3 Sensitivity Analysis for Personnel and Fire Brigade Response

The sensitivity discussed in this section is developed both qualitatively and quantitatively. To illustrate the process, the qualitative sensitivity assumes that personnel response effectively controls the fire on time. The quantitative sensitivity, in contrast, eliminates the credit to the fire brigade to highlight the importance of this fire protection feature.

A.9.3.1 Qualitative Evaluation

Figure A.11 highlights the impact of an increased *Manual Response* ranking from “Below Standard” to “Standard.” As described in Sect. A.7, the standard ranking results in the frequency being maintained through the success branch and a lower frequency for the failure branch to account for the “failure” of the event. The result is a reduced risk for the second and third scenarios when compared to the base case presented in Fig. A.3.

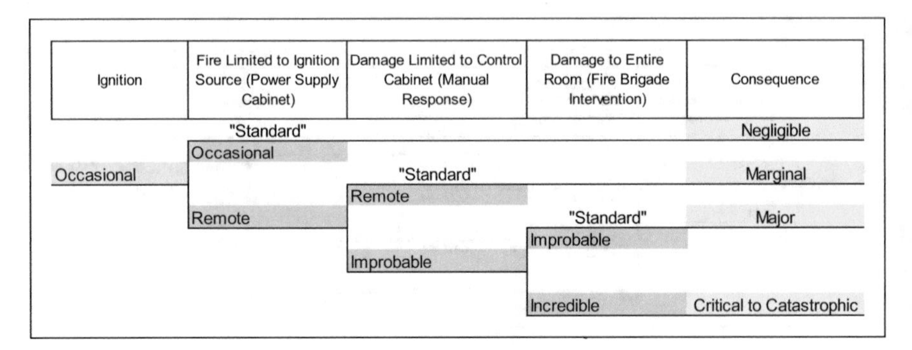

Fig. A.11 Scenario development event tree depicting “standard” manual response

A.9.3.2 Quantitative Evaluation

Manual suppression by the personnel or fire brigade is the only suppression means in the strategy. The analysis suggests a small margin (i.e., in the order of minutes) for the brigade to respond. Furthermore, the analysis assumed no brigade preparation or arrival time, further reducing the time available for suppression. Failure of the manual suppression response, as represented in Fig. A.12, suggests a relatively large risk increase. These results stress the need to monitor fire response times by facility personnel and the fire brigade.

Ignition	Fire Limited to Ignition Source (Power Supply Cabinet)	Damage Limited to Control Cabinet (Manual Response)	Sceanrio Frequency	Consequences		Risk
0.01	9.85E-01		9.85E-03	1.00E-05	Negligible	9.85E-08
	1.54E-02	0.0E+00	0.00E+00	1.00E-04	Marginal	0.00E+00
		0.0E+00	0.00E+00	1.00E-02	Major	0.00E+00
		1.0E+00	1.54E-04	1.00E+00	Critical to Catastrophic	1.54E-04

Fig. A.12 Scenario development event tree depicting the failure of manual suppression

A.9.4 Sensitivity Analysis for Automatic Fire Suppression

Solving the detection suppression event tree assuming an automatic suppression system with an availability of 0.98 suggests a significant risk reduction associated with this scenario. As expected, this is due to providing an effective alternative for suppression. An automatic system suggests risk values well below typical acceptability thresholds (Fig. A.13; Table A.9).

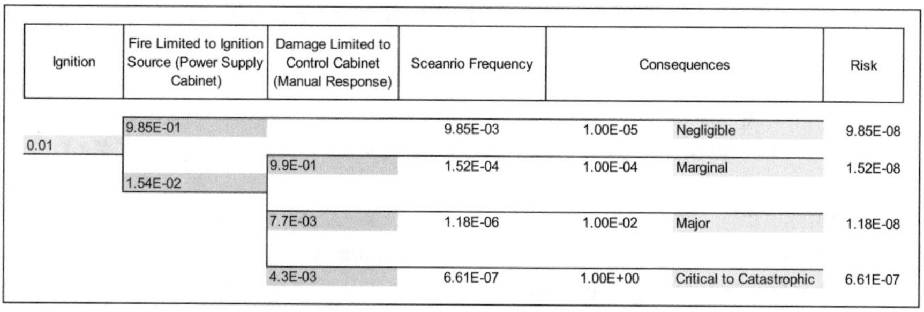

Ignition	Fire Limited to Ignition Source (Power Supply Cabinet)	Damage Limited to Control Cabinet (Manual Response)	Sceanrio Frequency	Consequences		Risk
0.01	9.85E-01		9.85E-03	1.00E-05	Negligible	9.85E-08
	1.54E-02	9.9E-01	1.52E-04	1.00E-04	Marginal	1.52E-08
		7.7E-03	1.18E-06	1.00E-02	Major	1.18E-08
		4.3E-03	6.61E-07	1.00E+00	Critical to Catastrophic	6.61E-07

Fig. A.13 Scenario development event tree depicting automatic suppression

Table A.9 Summary of conditional probabilities for detection and suppression reflecting automatic suppression system

Scenario	Elapsed time after fire ignition (min)	Estimated non-suppression probability P_{ns}	Cumulative suppression probability $P_s = 1 - P_{ns}$	Formula for the conditional probability of the scenario	Conditional probability of the scenario
1	10	0.012	0.988	$P_s(10)$	0.988
2	20	0.0043	0.9957	$P_s(20) - P_s(10)$	0.0077
3	>20			*Remaining probability*	0.0043

A.10 Example Summary

Similar to the example used in Chaps. 10 and 11 of this guide to describe the qualitative and quantitative risk assessment approach, this example provided a comprehensive overview of the recommended process for developing and documenting a fire risk assessment. In general, the following advantages and limitations associated with the example (and the guide in general) can be identified:

- The example is limited to the development of one scenario. It has been mentioned several times that the risk associated with a facility must include the contribution of all identified scenarios within the facility. From that perspective, the examples described in this guide are limited in scope. Often, risk acceptability or tolerability decisions are made on an individual scenario level and at a facility level, adding the risk of all the identified scenarios.

- Lack of data continues to be a challenge in developing a fire risk assessment. This guide attempted to provide practical alternatives to assess necessary values in cases where there is limited or no data. Specifically, the use of limited data supported by engineering judgment, analytical modeling, and uncertainty or sensitivity analyses are presented in this guide as alternatives to ensure those fire safety decisions are adequately justified.
- The example described a rigorous step-by-step approach illustrating how the different elements characterizing a fire protection program can be explicitly incorporated in the risk equation as necessary. This allows for the results of the fire risk assessment to be of practical use in the monitoring of key elements of the fire protection program throughout the operating life of the facility.
- The systematic approach described in this guide is applied to both the quantitative and qualitative assessment. This demonstrated consistency in both approaches in terms of risk insights and results.
- The inherent iterative process of the fire risk assessment is not practical to demonstrate in an example. As the risk of the scenarios identified in the facility is assessed and compared with each other, there will be a need to revisit some of the assessments to remove conservatisms (i.e., add realism). This will ensure that the resulting risk profile reflects the fire safety characteristics and day-to-day operation of the facility under study.

The risk assessment developed in this example suggested the following specific insights. The insights are based on risk results that do not have a large margin below the acceptability limits, although acceptable or tolerable. Furthermore, it is expected that including all the scenarios associated with this facility will further reduce the margin and increase the importance of these insights.

- Uncertainty in the fire ignition frequency indicates that strategies need to be implemented to ensure that these values are low. For example, establishing strict administrative procedures for controlling the presence of ignition sources and combustibles, routine inspection of equipment, and ensuring that the equipment is operating in appropriate environmental conditions may be necessary.
- Following up on the first insight listed above, fire prevention practices such as routine housekeeping are important. For example, ensuring safe distances/separation between ignition sources and combustibles (e.g., cables) may be necessary to prevent propagation.
- Manual fire suppression by facility personnel or fire brigade is the only alternative available. The analysis suggests that this requires a prompt response in the event of a fire. Routine training in response to alarms and fire events was identified as an important element to monitor.
- The installation of an automatic suppression system will provide an additional suppression capability and a significant risk reduction. There should be enough margin between the risk associated with the fire scenario and the acceptability limits with this reduction.

In summary, the fire risk assessment provided clear recommendations to maintain fire safety and potential improvements based on physical modifications.

References

US Regulatory Commission/EPRII, *EPRI/NRC-RES Fire PRA Methodology for Nuclear Power Facilities, Final Report*, NUREG/CR-6850, EPRI 1011989 (US Regulatory Commission/EPRII, Rockville, MD/Palo Alto, 2005)

U.S. Nuclear Regulatory Commission/EPRI, *Refining and Characterizing Heat Release Rates from Electrical Enclosures During Fire, Volume 1: Peak Heat Release Rates and Effect of Obstructed Plume, NUREG-2178*, vol 1 (U.S. Nuclear Regulatory Commission/EPRI, Washington, DC/Palo Alto, 2016)

U.S. Nuclear Regulatory Commission/EPRI, *Nuclear Power Plant Fire Ignition Frequency and Non-suppression Probability Estimation Using the Updated Fire Events Database: United States Fire Event Experience Through 2009, NUREG-2169* (U.S. Nuclear Regulatory Commission/EPRI, Washington, DC/Palo Alto, 2014)

SFPE Guide to Fire Risk Assessment, The Society of Fire Protection Engineers Series, https://doi.org/10.1007/978-3-031-17700-2

Index

SFPE Guide to Fire Risk Assessment, The Society of Fire Protection Engineers Series, https://doi.org/10.1007/978-3-031-17700-2

Q

R

S

T

U

Printed in the United States
by Baker & Taylor Publisher Services